T0199289

Advances in Swarm Intelligence for Optimizing Problems in Computer Science

Advances in Swarm Intelligence for Optimizing Problems in Computer Science

Edited by
Anand Nayyar, Dac-Nhuong Le,
and Nhu Gia Nguyen

CRC Press
Taylor & Francis Group
Boca Raton London New York

CRC Press is an imprint of the
Taylor & Francis Group, an **informa** business

CRC Press
Taylor & Francis Group

6000 Broken Sound Parkway NW, Suite 300
Boca Raton, FL 33487-2742

CRC Press is an imprint of Taylor & Francis Group, an Informa business

© 2019 by Taylor & Francis Group, LLC

No claim to original U.S. Government works

Printed on acid-free paper

International Standard Book Number-13: 978-1-138-48251-7 (Hardback)

Library of Congress Cataloging-in-Publication Data

Names: Nayyar, Anand, editor. | Le, Dac-Nhuong, 1983- editor. | Nguyen, Nhu Gia, editor.
Title: Advances in swarm intelligence for optimizing problems in computer science / Anand Nayyar, Dac-Nhuong Le, Nhu Gia Nguyen, editors.
Description: First edition. | Boca Raton,
FL : CRC Press/Taylor & Francis Group, [2019] | Includes bibliographical references and index.
Identifiers: LCCN 2018020199 | ISBN 9781138482517 (hardback : alk. paper) | ISBN 9780429445927 (ebook)
Subjects: LCSH: Swarm intelligence. | Computer algorithms.
Classification: LCC Q337.3 .A375 2019 | DDC 006.3/824–dc23
LC record available at https://lccn.loc.gov/2018020199

Typeset in Galliard
by Integra Software Services Pvt. Ltd.

Visit the Taylor & Francis Web site at
http://www.taylorandfrancis.com

and the CRC Press Web site at
http://www.crcpress.com

Printed and bound in Great Britain by
TJ International Ltd, Padstow, Cornwall

Contents

Preface

21st Century Globalization era's emerging technologies are driving various engineering firms to multifaceted states. The current challenges have demanded researchers to come up with innovative solutions to the problems. This basically has motivated the researchers to inherit ideas from Mother Nature and implement in various engineering areas, especially computer science. This way of implementation has led to the development of various biologically inspired algorithms that have all times proven to be highly efficient in handling all sorts of computational complex problems with high efficiency like Evolutionary Computation, Genetic Algorithm (GA), Ant Colony Optimization (ACO), Bee Colony Optimization (ABO), Particle Swarm Optimization (PSO), and many more.

Swarm Intelligence is basically the collection of nature inspired algorithms and is regarded as a concept under Evolutionary Computation. Researchers adapting Swarm Intelligence into different areas are concerned with the development of multiagent systems with applications like Optimization, Robotics Engineering, Network Improvement, Image Segmentation, and many more. The design parameters and approach for these systems are primarily different from other traditional methodologies.

The concept of Swarm Intelligence working principle is based on many unsophisticated entities that cooperate in order to exhibit a desired behaviour. The motivation behind the design and development of systems like this is based on the collective behaviour of social insects like Ants, Bees, Fishes, Fireflies, Elephants, Wasps, etc. Considering the individual members of the societies like Ants, Bees, etc., they are unsophisticated but with Mutual Cooperation and Coordinated behaviour, they achieve complex tasks.

Considering various insects, like Ants, Termites, and Wasps build highly sophisticated living colonies in cooperation, without any individuality but with a global master plan. The foraging behaviour helps the ants to search for food via chemical substances like Pheromone and determine the shortest path between source to next. Bee colonies are efficient in searching best food sources based on scouts that communicate information about the nearby food sources via Waggle Dance.

Scientists in various areas have applied Swarm Intelligence Principles in optimization and complex solutions building. Swarm Intelligence, in recent times, is concerned with collective behaviour in self-organized and decentralized systems. It was defined by Beni with regard to Cellular Robotic Systems, where simple agent organizes themselves via nearest neighbour interactions.

Motivated by exceptional capability of biologically inspired algorithms, the book will focus on various areas of Swarm Intelligence, its algorithms and applications cum implementations in areas of Computer Science.

We hope that the readers will enjoy reading this book and most importantly after reading this book, they see new areas to apply these algorithms with different approaches and perspectives.

Anand Nayyar
Dac-Nhuong Le
Nhu Gia Nguyen

Editors

Anand Nayyar received a PhD degree in Computer Science from Desh Bhagat University in Mandi Gobindgarh. Currently, he is on the Faculty as Researcher and Scientist at the Graduate School at Duy Tan University, Vietnam. He has 12 years of academic teaching experience with more than 250 publications in reputed international conferences, journals and book chapters (Indexed By: SCI, SCIE, Scopus, DBLP). Dr. Nayyar is a Certified Professional with more than 75 certifications from various IT companies like: CISCO, Microsoft, Oracle, Cyberoam, GAQM, Beingcert.com, ISQTB, EXIN, Google and many more. His areas of interest include: Wireless Sensor Networks, MANETS, Cloud Computing, Network Security, Swarm Intelligence, Machine Learning, Network Simulation, Ethical Hacking, Forensics, Internet of Things (IoT), Big Data, Linux and Open Source and Next Generation Wireless Communications. In addition, Dr. Nayyar is a Programme Committee Member/Technical Committee Member/Reviewer for more than 300 international conferences to date. He has published 18 books in Computer Science by GRIN, Scholar Press, VSRD Publishing. He has received 20 Awards for Teaching and Research including: Young Scientist, Best Scientist, Exemplary Educationist, Young Researcher and Outstanding Reviewer Award.

Dac-Nhuong Le, PhD, is Deputy-Head of Faculty for Information Technology and Vice-Director of the Information Technology Apply Center at Haiphong University, Vietnam. Dr. Le is also a research scientist at the Research and Development Center of Visualization & Simulation (CSV) at Duy Tan University in Danang, Vietnam. He has more than 45 publications in reputed international conferences, journals, and online book chapter contributions (Indexed By: SCI, SCIE, SSCI, Scopus, DBLP). His areas of research include: Evaluation Computing and Approximate Algorithms, Network Communication, Security and Vulnerability, Network Performance Analysis and Simulation, Cloud Computing, plus Image Processing in Biomedical. His core work is in Network Security, Wireless, Soft Computing, Mobile Computing and Biomedical. Recently, Dr. Le has been on the Technique Program Committee and the Technique Reviews, as well as being the Track Chair for the following international conferences: FICTA 2014, CSI 2014, IC4SD 2015, ICICT 2015, INDIA 2015, IC3T 2015, INDIA 2016, FICTA 2016, IC3T 2016, ICDECT 2016, IUKM 2016, INDIA 2017, FICTA 2017, CISC 2017, ICICC 2018, ICCUT 2018 under Springer-ASIC/LNAI/CISC Series. Presently, Dr. Le is serving on the editorial board of several international journals and he has authored six computer science books by Springer, Wiley, CRC Press, Lambert Publication, VSRD Academic Publishing and Scholar Press.

Nhu Gia Nguyen received a PhD degree in Computer Science from Ha Noi University of Science at Vietnam National University, Vietnam. Currently, Dr. Nguyen is Dean of the Graduate School at Duy Tan University, Vietnam. He has 19 years of academic teaching experience with more than 50 publications in reputed international conferences, journals, and online book chapter contributions (Indexed By: SCI, SCIE, SSCI, Scopus, DBLP). His areas of research include: Network Communication, Security and Vulnerability, Network Performance Analysis and Simulation, Cloud Computing, Image Processing in Biomedical. Recently, Dr. Nguyen has been on the Technique Program Committee and the Technique Reviews, in addition to being the Track Chair for the following international conferences: FICTA 2014, ICICT 2015, INDIA 2015, IC3T 2015, INDIA 2016, FICTA 2016, IC3T 2016, IUKM 2016, INDIA 2017, INISCOM 2018, FICTA 2018 under the Springer-ASIC/LNAI Series. Presently, he is Associate Editor of the International Journal of Synthetic Emotions (IJSE).

Contributors

M. Balamurugan
School of Electrical Engineering
VIT University
Vellore, India

Surbhi Garg
Bharati Vidyapeeth's
    Institute of Computer
    Applications and Management
New Delhi, India

Dr. Deepak Gupta
Department of Computer Science
    and Engineering
Maharaja Agrasen Institute of
    Technology
New Delhi, India

Sanjay Jain
Department of Computer Science
    and Engineering
Amity University
Rajasthan, India

Dr. Ashish Khanna
Department of Computer Science
    and Engineering
Maharaja Agrasen
    Institute of Technology
New Delhi, India

Dr. Sandeep Kumar
Department of Computer Science
    and Engineering
Amity University
Rajasthan, India

Dr. Akshi Kumar
Department of
    Computer Science & Engineering
Delhi Technological
    University
New Delhi, India

Dr. Rajani Kumari
Department of IT & CA
JECRC University
Jaipur, India

Bandana Mahapatra
Department of Computer Science
    and Engineering
SOA University
Bhubaneshwar, India

D. Komagal Meenakshi
Department of CSE, CMRIT
Visvesvaraya Technological
    University
Karnataka, India

S. Narendiran
Department of Instrumentation
    and Control Engineering
Sri Krishna College
    of Technology
Coimbatore, India

Dr. Nhu Gia Nguyen
Graduate School
Duy Tan University
Da Nang, Vietnam

Dr. Anand Nayyar
Graduate School
Duy Tan University
Da Nang, Vietnam

Prof. Srikanta Patnaik
Department of
  Computer Science and
  Engineering
SOA University
Bhubaneshwar, India

Dr. Prem Kumar Ramesh
Department of CSE, CMRIT
Visvesvaraya Technological
  University
Karnataka, India

Dr. Sarat Kumar Sahoo
Department of
  Electrical Engineering
Parala Maharaja
  Engineering College
Berhampur, Odisha

Shanthi M. B.
Department of CSE, CMRIT
Visvesvaraya Technological
  University
Karnataka, India

Dr. Harish Sharma
Department of Computer Science
  and Engineering
Rajasthan Technical University
Kota, Rajasthan

K. Vikram
Department of Information
  Technology MPSTME
NMIMS University
Mumbai, India

1

Evolutionary Computation

Theory and Algorithms

Anand Nayyar, Surbhi Garg, Deepak Gupta, and Ashish Khanna

CONTENTS

1.1 History of Evolutionary Computation

As said earlier, history is always accountable for the present or the current developments. Hence, this necessitates us knowing and appreciating the precious contributions of the researchers from the past in the field of **evolutionary computation.** Evolutionary computation is a thrilling field of computer science that focuses on Darwin's principle of natural selection for building, studying, and implementing algorithms to solve various research problems. The concept of implementing the **Darwin Principle** to solve automated problems evolved from the late 1940s, way before the succession of computers (Fogel, 1998). Darwin's theory states 'the fittest individuals reproduce and survive'. In year 1948, '**genetic and evolutionary search**' was initiated by Turing in his PhD thesis, where he developed the concept of an automatic machine, presently known as 'Turing Machine', which was then an unorganized machine to execute certain experiments that later gained popularity as the 'Genetic Algorithm.' It was practically executed as computer experiments on '**optimization through evolution and recombination**' by Bremermann in 1962. Apart from this, three variations of this basic idea were proposed during the 1960s. Fogel, Owens, and Walsh further extended it toward evolutionary programing and named it **genetic algorithms** (De Jong, 1975; Holland, 1973) in Holland, whereas Rechenberg and Schwefel contributed the invention of **evolutionary strategies** (Rechenberg, 1973; Schwefel, 1995) in Germany. These areas evolved separately for a period of 15 years in their respective countries until the early 1990s, after which they all together were named **evolutionary computation** (EC) (Bäck, 1996; Bäck, Fogel, & Michalewicz, 2000). In the early 1990s the **genetic programing** concept came up as a fourth stream, following the general ideas upheld by Koza (Banzhaf, Nordin, Keller, & Francone, 1998; Koza, 1992). The scientific forums declared EC to keep a record of the past as well as future events conducted. The International Conference on Genetic Algorithms (ICGA) was first conducted in 1985 and was repeated every second year until 1997; it merged with the annual Conference on Genetic Programming in 1999 and was renamed the Genetic and Evolutionary Computation Conference (GECCO). The Annual Conference on Evolutionary Programming (EP), which has been held since 1992, merged with the IEEE Congress of Evolutionary Computation (CEC) conference, held annually since 1994. The first European event to address all the streams, named Parallel Problem Solving from Nature (PPSN), was held in 1990. *Evolutionary Computing*, said to be the first scientific journal, launched in 1993. In 1997, the European

Commission implemented the European Research Network in Evolutionary Computing, named it EVONET, and funded it until 2003. Three major conferences, CEC, GECCO, PPSN, and many small conferences were held dedicated to theoretical analysis and development. The Foundation of Genetic Algorithms has existed since 1990 along with EVONET, the annual conference. Roughly over 2000 evolutionary computation publications were done by 2014, which includes journals as well as conference proceedings related to specific applications.

1.2 Motivation via Biological Evidence

The Darwinian theory of evolution and insight into genetics together explain the dynamics behind the emergence of life on Earth. This section covers Darwin's theory of survival of the fittest and genetics in detail.

1.2.1 Darwinian Evolution

The origins of biological diversification and its underlying mechanisms have been well captured by Darwin's theory of evolution (Darwin, 1859). The underlying principle states that natural selection is always positive toward the individual that challenges others to acquire a given resource most effectively. That is, who adapts best to the given environment. This phenomenon is widely known as **survival of the fittest**. The theory follows the concept of 'The excellent will flourish, and the rest will perish'. Competition-based solution is the pillar on which evolutionary computation rests. The other primary force identified by Darwin is the phenotypic variations among the members of the population, which includes the behaviour and physical features of the individual that act as the force behind its reaction to the environment. Each individual contains a phenotypic trait that occurs in unique combinations, evaluated by environment. The individual has a higher chance of creating offspring if the combination provides favorable values; otherwise, it suffers rejection without producing any offspring. Moreover, inheritable favorable phenotypic traits might propagate from the individual through its offspring. Darwin's insight includes **random variation, or mutation**, in phenotypic traits taking place from one generation to the next, which gives birth to new combinations of traits. Here, the best one survives for further reproduction, contributing to the process of evolution.

This process described in Darwin's theory is captured in the adaptive landscape, called the 'adaptive surface' (Wright, 1932). The surface has a height dimension corresponding to the fitness, where high altitude shows high fitness. The other two or more dimensions correspond to the biological traits, where the x-y plane shows all combinations of traits possible, and z values correspond to the fitness. Each of the peak signifies the possible combination of traits. Hence, evolution can be considered as the gradual advance of the population to the high-altitude areas influenced by factors like variation and natural selection.

Our acquaintance with physical landscape takes us toward the concept of multimodal problems, where there are a number of points that are better in comparison to the existing neighbouring solutions. All these points are called as **local optimal**, whereas the highest point among them is the **global optimum. Unimodal** is the name given to a problem with a single local optimum. The population being a set, with random choices being made at selection, and variation operators, gives rise to genetic drift, whereby highly fit individuals may be lost, followed by loss in verity. This phenomenon is termed a 'meltdown' of the hill, entering the low-fitness valleys.

The combined effects of the drift and selection enable the population to move uphill as well as downhill.

1.2.2 Genetics

Molecular genetics provides a microscopic view of a natural evolution. It crosses the general visible phenotypic features, going deeper in the process. The key observation in genetics is the individual being a dual entity. The external feature is **phenotypic** properties, which are constructed at the level, i.e., internal construction. Genes here may be considered as the functional units of inheritance, encoding the phenotypic characters, i.e., the external factors visible, e.g., fur colour, trail length, etc. Genes may hold many properties from the possible alleles. An allele is one of the possible values a gene can have. Hence, an allele can be said to have a value that a variable can have mathematically. In a natural system, a single gene may affect many phenotypic traits, which is called **pleiotropy**. In turn, one phenotypic trait can be determined to be the result of a combination of many genes, termed 'polygene.' Hence, biologically the phenotypic variations are connected to the genotypic variations, which are actually an outcome of gene mutation, or the recombination of genes by sexual reproduction.

A genotype in general consists of all information required to build a specific phenotype. The genome contains the complete genetic information of any living organism with its total building plan. All the genes of an organism are arranged in several chromosomes. The human body contains 46 chromosomes; the other higher life forms like plants and animals also contain double complemented chromosomes in their cells. These cells are called **diploid**. The chromosomes in human diploid cells are organized into 23 chromosomal pairs. The gametes (i.e., sperm or egg cell) are haploid, containing only a single complement of a chromosome. The haploid sperm cell merges with egg cell, forming a diploid cell or a

zygote. Within a zygote, the chromosome pair is formed out of the paternal and the maternal half. The new organism formed by the zygote is called as the **ontogenesis** and keeps intact the genetic information of the cell.

The combination of specific traits from two individuals into an offspring is called a crossover, which occurs during the process of **meiosis**.

As we know, all living organisms are based upon DNA, which is structurally composed of double helix of nucleotides encoding the characteristics of a whole organism. The triplets of nucleotides are called **codons**, each of which codes for a specific amino acid. Genes can be described as larger structures based upon DNA, containing many codons, carrying the codes of protein. The transformation of DNA to protein consists of two main steps: transcription, where information of DNA is written to RNA, and translation, the step creating a protein from the RNA. This perspective confirms that changes in the genetic material of a population can only happen from random variation and the natural selection process, not from individual learning. The important concept here is all variation occurs at the genotypic level, whereas the selection is based upon actual performance in a considered environment.

1.3 Why Evolutionary Computing?

Evolutionary computing in essence refers to a set of techniques obtained from the theory of biological evolution used to make streamlined techniques or procedures, typically executed on PCs that are utilized to tackle issues.

Mathematics and computer sciences formulate the base for the evolution of the algorithm. Nature's solutions have always served as a source of inspiration in the engineering domain. Natural problem solvers may very well form the basis of automatic problem solving.

Two things that strike the mind, when it comes to natural problem solving, are:

- The human brain (the inventor of things).
- The evolutionary process (which created the human brain).

Attempting to outline problem solvers in view of these factors forms the basis of neurocomputing and evolutionary computing respectively (Eiben & Smith, 2007a).

Artificial intelligence aims at modeling the human brain's techniques of processing information. To manufacture canny systems that model intelligent behaviour is the task at hand. Evolutionary computing differs from the traditional searching and optimizing in the following ways:

- Utilizing potential solutions to aid the searching process.
- Rather than using function derivatives or related knowledge, it uses direct 'fitness' information.
- Using probabilistic, as opposed to deterministic, transition guidelines (Eberhart, Shi, & Kennedy, 2001).

Research and development are not keeping up with the pace of increasing demand for solution automation. As a result, the time for analyzing and tailoring algorithms is decreasing. In parallel to less time, the complexity of problems is increasing. These two patterns suggest an earnest requirement for robust algorithms with satisfactory performance and adherence to the constraints of limited capacity. This suggests a need for customizable solutions for specific issues that are applicable on a variety of problems and deliver good solutions in adequate time. Evolutionary algorithms cater to these needs and thus answer the challenge of making automated solution methods for a large number of problems, which are increasingly mind-boggling, in the least possible time (Eiben & Smith, 2007a).

As said above, evolutionary computing gives good solutions, implying that the solutions are not necessarily optimal – thus, making us think, why should we be satisfied with them when we have traditional approaches that may give optimal solutions? Reasons for this paradigm shift are as follows:

- Considering feasibility as a parameter, traditional and customized approaches will not be feasible frequently. On the contrary, evolutionary computing paradigms will be usable in a vast number of situations.
- The real strength of evolutionary computing paradigms lies in their robustness. In this context, robustness means that an algorithm can be used to solve numerous problems and a wide variety of problems, with a minimum amount of special acclimations to represent exceptional characteristics of a specific problem.

This brings us to our *Law of Sufficiency*: If a solution is *good* enough, and it is *fast* enough, and it is *cheap* enough, then it is *sufficient*. In nearly all real-world applications, we are searching for, and we are happy with adequate arrangements. Here 'adequate' implies that the solution meets determinations.

1.4 Concept of Evolutionary Algorithms

According to Technopedia, 'An evolutionary algorithm is considered a component of evolutionary computation in artificial intelligence. An evolutionary algorithm functions through the selection process in which the least fit members of the population set are eliminated, whereas the fit members are allowed to survive and continue until better solutions are determined.'

https://www.technopedia.com/definition/32751/
evolutionary-algorithm

In other words,

'Evolutionary algorithms are computer applications that mimic biological processes in order to solve complex problems.'

Despite the fact that we have numerous evolutionary algorithms, the common idea behind all of them remains the same: given a population of individuals, the environmental process causes common selection (survival of the fittest), and this causes a spike in the fitness of population. Arbitrary sets of possible solutions can be created, i.e., elements belonging to function domain applied to a quality function as an abstract fitness measure – the higher the better with respect to a given quality function to be maximized. Better candidates are chosen on the basis of obtained fitness to seed the next generation by applying recombination and/or mutation to them. Recombination refers to an operator applied on two or more selected candidates (or parents), resulting in one or more new candidates (or children). However, mutation is applied to a single candidate, resulting in a new candidate. Executing them together prompts a set of new applicants (the offspring) that contend in light of their fitness (and potentially age) with the old ones for a place in the new generation. This process can be repeated until a solution, i.e., a candidate with sufficient quality, is found or a previously set computation limit is touched.

Concluding the process, there are two fundamental factors forming the basis of evolutionary systems:

- Variation operators (recombination and mutation), which create necessary diversity and in effect facilitate novelty.
- Selection, which acts as a force pushing quality.

Improving fitness values in next generations on the application of variation and selection may give an illusionary effect of optimizing or at least approximating optimization. But, since the choice of candidates (parents in the case of recombination and one candidate in the case of mutation) is arbitrary, even the weak individuals may survive (chances are lower than for fitter individuals). The following Figure 1.1 demonstrates the general flow of an Evolutionary Algorithm.

An alternate view of evolution is the process of adaptation. Following this view, fitness is seen as an expression of environmental requirements rather than a function to be optimized.

Evolutionary algorithms encompass many features, on the basis of which they can be positioned in the family of 'generate and test' algorithms:

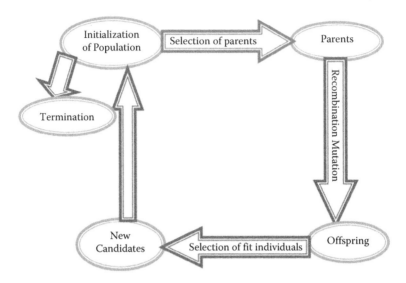

FIGURE 1.1
Flow chart representing general flow of an evolutionary algorithm.

- They are population based, as they simultaneously process a collection of possible solutions.
- They mostly use recombination to form a new solution by mixing the information of previous candidates.
- They are stochastic (Eiben & Smith, 2007b).

Having said that, evolutionary computational techniques have a set of common features, but they also have distinctive features that can be seen by digging into the level of abstraction. They may differ in terms of data structure used and consequently the 'genetic' operators. Even a particular technique may exhibit various forms and twists. For example, selection of individuals can be done by applying numerous approaches such as:

- Proportional selection, in which the chances of selection increase with the fitness of an individual, implying the probability of selection is proportional to fitness. Detailing further, it may require use of scaling windows or truncation methods.
- The ranking method, in which individuals are ranked on a scale of best to worst with a fixed probability of selection for complete process of evolution. Moreover, probability may be a linear or non-linear representation.

- Tournament selection, where the next generation is selected on the basis of competition of a number of individuals (generally two), and the competition step is iterated a population size number of times. Tournament selection methods depend largely on the size of tournaments.

Likewise, a generational policy can be selected deterministically or non-deterministically from these alternatives: replacing the complete population with offspring, to choose the best individuals from two generations, steady-state systems in which an offspring is produced that will replace possibly the worst, or an elitist model in which the best individuals are carried from one generation to other successive generations.

Data structures along with a set of genetic operators, being the most important parts of an evolutionary algorithm, form the basis of distinction between various paradigms of evolutionary algorithms (Michalewicz, Hinterding, & Michalewicz, 1997).

1.5 Components of Evolutionary Algorithms

As per the unified approach suggested by De Jong (1975), evolutionary algorithms can be designed on the basis of the following components: representation, fitness function, parent's selection, population, crossover operators, mutation operators, survival selection, and terminal condition (Wong, 2015).

1.5.1 Representation

The initial step of defining an evolutionary algorithm is to bridge the gap between 'the real world' (or the original problem) and 'the EA world' (problem solving space where evolution will take place). Objects shaping conceivable solutions in the original problem context are alluded to as phenotypes; their encoding, the individuals, within the EA are called genotypes (Eiben & Smith, 2007b).

This step includes genotype-phenotype mapping. For example, if integers are phenotypes, we can choose to represent them as binary in phenotypes, say, 11 is represented as 1011 and 54 is represented as 110110. If we mutate the first bit, we get 3(0011) and 22(010110). From this example, we can note that mutating a single bit in the genotype may vary from phenotype. Generally, genotype representations are kept compact and close to phenotype representations such that there is no loss of semantic information while mapping genotype to phenotype for measurement metrics, say distance. There are a variety of representations that can be adopted by an evolutionary algorithm, for example, fixed length linear, variable length-linear, or tree structures (Wong, 2015).

Elements of genotype space and phenotype space have various synonyms under common EC terminology. Phenotype space, the space having points of possible solutions can be denoted by candidate solution, phenotype and individual whereas genotype space, space having points of evolutionary search can be represented by chromosome, genotype, and individual. Similarly, elements of individuals such as variable, position, or object are synonymous to placeholder, locus, or gene, and value or an allele, respectively.

Representations are not only about mapping from phenotype to genotype, they must also be invertible, implying that for each genotype there should be at most one corresponding phenotype. Other than mapping, representation may also emphasize 'data structures' of the genotype space (Eiben & Smith, 2007b).

1.5.2 Fitness Function

This function represents requirements to be adopted, forms the basis of selection, and opens the door for improvement. It also represents the task to be solved in an evolutionary context. Ordinarily, this function is formed from a quality measure in the phenotype space and the inverse portrayal. For example, to maximize x^2 on integers, the fitness of genotype 1011 could be defined as the square of its phenotype, i.e., $11^2 = 121$. This function is also termed an evaluation function. However, the problem may require minimization for fitness, which is generally associated with maximization, and changing minimization to maximization and vice versa is not child's play in terms of mathematics. If the goal of the problem is optimization, the function is termed as objective function in the context of the original problem, and the evaluation function is either identical or simple transformation of the same (Eiben & Smith, 2007b).

1.5.3 Parents' Selection

Selection of parents for the next generation is an important step, and the parent selection process selects 'better' individuals to serve as parents for the next generation, where x is said to be better than y, if x has a higher fitness than y. An individual is a parent if it is involved in the creation of offspring. Parent selection adds quality measures to the process in the sense that better parents will generate better offspring. Due to the probabilistic nature of the selection process, high-quality individuals have an upper hand, but low-quality individuals also have a small but positive chance, thus avoiding the search to adapt an excessively greedy approach or to get restricted to local optimum (Eiben & Smith, 2007b; Wong, 2015).

1.5.4 Population

Population is a set of individuals, individuals which are possible solutions in reference to a given problem. A population may also be referred to as a set allowing multiple copies of an individual, i.e., a multiset of genotypes. Individuals are constant, but the population is changing. The population size refers to the number of individuals in it. An additional space, with neighbourhood relation or distance measure can also be found in some highly developed evolutionary algorithms, and to fully specify population in such cases, an additional structure also needs to be defined. Unlike variation operators that work on at most two parent individuals, selection operators (parent selection and survivor selection) work at population level, i.e., best individual of the given set for the next population, and worst individual of the set to be replaced by a new one. During the evolutionary search, population size remains constant in almost all applications of evolutionary algorithms. Solutions may differ from each other in terms of fitness values, different phenotypes, different genotypes, or statistical measures such as entropy, consequently diversifying the population. An important point to note is that one fitness value does not imply one phenotype, and further one phenotype does not imply one genotype. But the converse of this statement is not true, i.e., one genotype can map back to only one phenotype and in turn one fitness value (Eiben & Smith, 2007b).

1.5.5 Crossover Operators or Recombination Operators

Crossover or recombination operators take after the regeneration process in nature. Combined with mutation operators, they are collectively termed reproductive operators. All in all, a crossover operator consolidates two individuals to frame a new individual. The operator performs a two-step process of partitioning an individual and then assembling parts of two individuals into a new one. The partitioning relies upon the representation embraced and is not too trivial. In this way, it can be envisioned that these operators are not representation dependent.

Classic crossover operators are:

- One-Point Crossover: Widely used operator in this category, because of its simple nature. A cut point in genomes is randomly chosen in the given two individuals, and then the parts after (or before) the cut points are swapped between two genomes.
- Two-Points Crossover: To avoid positional bias that may occur in one-point crossover, as only leftmost or rightmost parts are swapped in it, a two-points crossover is used.
- Uniform Crossover: Each gene is assigned an equal probability to be swapped.

- Blend Crossover: Commonly used in optimizing real numbers. Rather than swapping genes, it tries to blend the two together with the help of arithmetic averaging to obtain intermediate values (Wong, 2015).

It is a binary variation operator and is stochastic in nature as the choice of parts of parents to be combined is random. Recombination operators using more than two parents are easy to implement but have no biological equivalent. This may be a reason behind their not so common use, despite the fact that they have positive impact on evolution as indicated by several studies. The idea behind the working is simple – to obtain an offspring with some desirable features, we mate two individuals with different yet required features. The process being random, this may also lead to an undesirable combination of traits (Jones, 2002).

1.5.6 Mutation Operators

Mutation operators modify the parts of a genome of an individual stochastically. In other contexts, they can be thought as an investigation component to balance the exploitation power of crossover operators. They are representation dependent (Wong, 2015).

Mutation operators work on one genotype and result into an offspring with some modifications. The generated mutant (child or offspring) is dependent on the results of a series of random choices. Any unary operator can't be termed as mutation; an operator working on an individual with respect to a specific problem is called a mutation operator. Talking about evolutionary computing, mutation may play several roles; for example, it is not used in genetic programming; on the contrary, it is used in genetic algorithms as a background operator supporting regeneration by filling the gene pool with 'fresh blood' and in evolutionary programming, it is the only variation operator for completing the search process (Eiben & Smith, 2007b).

Commonly used mutation operators are:

- Bitflip Mutation: Each bit in a binary genome is inverted with a predefined probability.
- Random Mutation: A generalized version of bitflip mutation, it can be applied on multiple genomes; each part of the genome is replaced by any random value within a problem-specific domain. This replacement process is governed by a predefined set of probabilities.
- Delta Mutation: Generally used in real number genomes; each number is incremented or decremented by a number. This value by which a real number is incremented is termed as delta. Generally, the probability of change and delta is predefined.

- Gaussian Mutation: This is similar to delta mutation with the only difference in the value of delta, which should be a Gaussian random number here (Wong, 2015).

1.5.7 Survivor Selection

This is a process similar to that of parent selection but occurs on a different stage. This process comes into the picture when we have the offspring and we have to decide which individuals are to be carried into the next generation. Selection is generally fitness based, but age is also used in some of the algorithms. Opposite to the non-deterministic nature of parent's selection, this is generally deterministic.

Many times, this process is also called replacement, but if we consider n children out of a population having size m, where n < m, it is possible to say that we need to find n candidates to be replaced, unless all replaced are the survivors; on the contrary, if n > m, this is not the case. That's why the term survivor selection is an apt choice (Eiben & Smith, 2007b).

1.5.8 Termination Condition

This refers to the stage when algorithms should stop. The optimum solution is always best if we achieve it, but due to the random nature of evolutionary algorithms, optimality can't be guaranteed.

Therefore, some other considerations are:

- Maximum allotted CPU time pass-by.
- As fitness evolutions can be expensive to compute in some cases, reaching a given limit set on the total number of evaluations.
- If for many successive generations, improvement in fitness remains under a threshold value.
- Diversity of population downfalls under a threshold (Eiben & Smith, 2007b).
- When a certain number of generations is reached (Wong, 2015).

Thus, termination will be either at optimal solution or by satisfying any of the mentioned conditions (Eiben & Smith, 2007b).

1.6 Working of Evolutionary Algorithms

Algorithms under this category have some common features relevant to their working. Let's talk about the working principle with the help of an example, such as a single dimensional function, which is to be maximized. Individuals may undergo many changes in the search process summarized as:

- Initial stage: Individuals are covering complete search space in a random order.
- Mid-stage: Individuals start concentrating toward peaks, as a result of selection and variation operators.
- Last stage: All individuals are accumulated over some peaks.

The following Figure 1.2 represents initial stage and fluctuations in graph can be observed in Figure 1.3 after some generations.

At the end, there is a possibility that none of the individual has reached global optimum. 'As mentioned by Eiben Smith, distinct phases of search are categorized in terms of exploration (the generation of individuals in as-yet untested regions of test space) and exploitation (the concentration of search in the vicinity of known good solutions).' Losing population diversity at an early stage is called premature convergence, which may result in a trap of the local optimum. This problem is very commonly found in evolutionary algorithms. Another effect is any time behaviour. An algorithm may be increasingly progressive in the beginning but may achieve stability after some time. This effect is often found in algorithms working on re-enhancements. The term 'any time' represents the flexibility to stop the search process at any time, irrespective of optimality of solution. A question that follows this is, will it be more fruitful if we imply some intelligent ways to seed the initial population rather than doing it randomly?

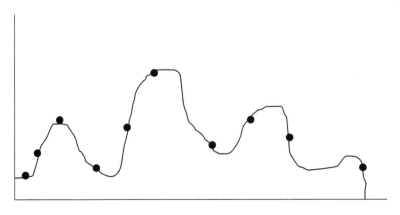

FIGURE 1.2
Initial stage.

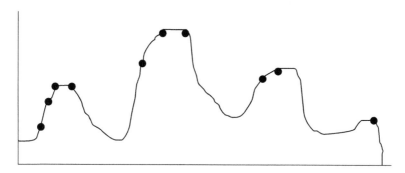

FIGURE 1.3
After some generations.

- At the initialization level, if initialization is done by applying intelligent ways (Stochastic, Deterministic Methods), the search will have better population to start with, but this can be reached after some generations in the case of random initialization: thus, the gain is doubtful.

- At termination level, if we divide runs into two equal parts, the increase in fitness will be more significant in the first half than the second, thus indicating there is no profit in very long runs because of any time behaviour.

Taking into account the performance of evolutionary algorithms on a wide range of problems, their performance is good over a variety of problems. According to the 1980s view, evolutionary algorithms are better than random search because their performance doesn't degrade suddenly with problem types and customizations, but a random search algorithm is efficient in the case of a specific problem only. Thus, these two are diametrically opposite. In the modern view, these two extremes can be combined to form a hybrid algorithm. It may be incorrect to say that evolutionary algorithms always give a higher performance than random search algorithms, according to the 'No Free Lunch Theorem', contradicting the 1980s view of evolutionary algorithm as shown in Figure 1.4 (Eiben & Smith, 2007b).

1.7 Evolutionary Computation Techniques and Paradigms

1.7.1 Genetic Algorithms

As mentioned in Banzhaf et al. (1998), 'Genetic Algorithm is inspired by the mechanism of natural selection, a biological process in which stronger individuals are likely to be the winners in a competing environment.' John

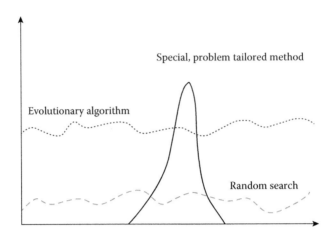

FIGURE 1.4
1980s' view of evolutionary algorithm.

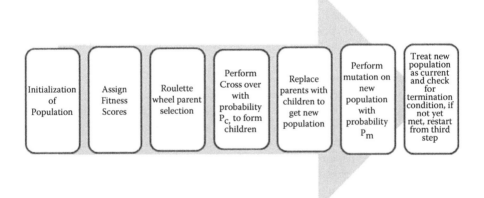

FIGURE 1.5
Genetic algorithm steps developed by Holland.

Holland (University of Michigan) invented them in the 1960s. A genetic algorithm performs multi-directional search. The reason behind this feature is a set of possible solutions (population) and information generation and exchange across directions. They represent adaptation and operate on binary strings (Michalewicz et al., 1997). Genetic Algorithm proposed by Holland is presented in Figure 1.5.

Let's understand the process with the following example:

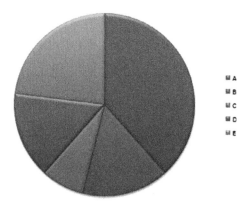

FIGURE 1.6
Roulette wheel parent selection.

Let's consider Roulette Wheel Parent Selection Pie Diagram in Figure 1.6. Suppose, there are five individuals in a population having fitness as follows:

Individual	Fitness
A	5
B	2
C	1
D	2
E	3

The roulette wheel is divided into five sectors; each sector corresponds to an individual, and the area of sector is proportional to its fitness. Parent selection is done by spinning the wheel. The selection being random, the wheel can choose any individual as parent, which is undoubtedly in favor of fitter individuals. Parents need not be unique, and in every cycle, fit individuals may give rise to numerous offspring (Ergin, Uyar, & Yayimli, 2011).

n/2 pairs are selected out of n individuals as parents, forming a new population. To obtain a pair of offspring, a crossover operator with probability P_c is applied on each parent pair. A random cross point is selected. A child reflects properties of a parent before that point and the other after that point. Figure 1.7(a) represents pair of parents, i.e., initial population chosen and Figure 1.7(b) shows the resultant offsprings that can be generated.

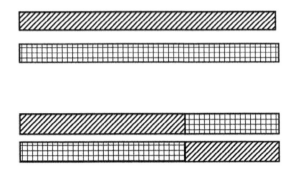

FIGURE 1.7
(a) Pair of parents; (b) Pair of offspring.

Now, offspring becomes part of the population, replacing their parents. Mutation is applied on them, i.e., bit inversion of all bits takes place with probability P_m. P_m and P_c (probability of mutation and probability of crossover, respectively) are customizable parameters of the algorithm. General values are 0.01 and 0.8, respectively indicating the greater importance of crossover and mutation as a background operator (Jones, 2002).

Nowadays genetic algorithms and evolutionary algorithms are used analogously (Wong, 2015).

Every problem is represented as a binary string. For example, in z = (1001010101011100), the first half of the string represents the x coordinate values and the second half represents y coordinate values. Recombination operators and mutation operators are used to complete the process. Mutation, as written above, changes a bit and crossover interchanges the bits of two parents. For instance, if parents are (00110) and (10101) a crossover result could be (00101) and (10110) (crossing after second bit) (Michalewicz et al., 1997).

However, this representation is not well suited for several problems. Thus, some other representations, such as an integer, strings, a tree, and a matrix are used by many algorithms. Different encodings require specially designed crossover operators. To increase effectiveness, hybrid genetic algorithms combine specialized operations with two other operators (Jones, 2002).

Similar to a child playing with building blocks, in which a big building is built by joining small blocks, a genetic algorithm finds a close to optimal solution, through short, low-order and high-performance schemata (Michalewicz et al., 1997).

An issue with widely accepted genetic algorithms is that the survival of good individuals is not guaranteed after every cycle. A new population is not produced by all algorithms after each cycle. It would be better to replace the worst individuals and keep the best individuals of the old population unchanged to form a new population (Man, Tang, & Kwong, 1996).

1.7.2 Evolution Strategies

These were developed as methods to solve parameterized optimization problems. An individual is represented as a vector, i.e., $\vec{v} = (\vec{x}, \vec{\sigma})$ having float values. This idea of representation in terms of float values is not found in genetic algorithms. In the notation, $\vec{x}$ denotes any point in the search space and $\vec{\sigma}$ denotes the standard deviation. Mutation is computed as:

$\vec{x}^{t+1} = \vec{x}^{t} + N(0, \vec{\sigma})$, Where $N(0, \vec{\sigma})$ is a vector representing independent random Gaussian numbers with the mean as 0 and standard deviation as $\vec{\sigma}$. The resultant offspring replaces its parent in the case where it has better fitness and fulfills the required criteria; otherwise, it is discarded. This idea is based on the fact that small changes are more frequent than large changes. The standard deviation vector is constant during the entire process (Michalewicz et al., 1997).

Further evolutionary strategies changed to add controlling parameters, which affects the behaviour of the algorithm. This is denoted by an interesting notation: $\left(\frac{\mu}{\rho+}, \lambda\right) - ES$ where μ denotes parent population size, ρ denotes breeding size, and $\left(\frac{\mu}{\rho} + \lambda\right) - ES$ denotes the algorithm is overlapping; not overlapping is denoted by $\left(\frac{\mu}{\rho}, \lambda\right)$. λ denotes offspring population size (Wong, 2015).

The advantage over the previous notation is that the vector representing standard notation is neither constant nor is changed by a deterministic algorithm. ES denotes evolution strategy.

1.7.3 Evolutionary Programming

Evolutionary programming was developed by Lawrence Fogel and is aimed at the development of artificial intelligence in the field of ability to anticipate changes in a domain. The domain is represented as finite alphabet sequence, and a new symbol is expected as an output of the algorithm. A measure of accuracy of prediction, called the pay-off function, should be maximum with the output produced by algorithm.

The individual's chromosomal representation is done by a finite state machine (FSM) because it can model the behaviour on the basis of symbol interpretation. An FSM consists of possible states and arrows representing change of state from one state to another. Symbols above the edge are represented as input/output. Thus, a population is represented by a set of finite state machines, and each individual is a possible solution.

The following Figure 1.8 demonstrates finite state machine of parity checker.

The machine starts at even, and goes to odd only when the input 1 is consumed, leading to parity bit as 1. Numbers of predictions are equal to the number of inputs scanned, and fitness is calculated with the help of the accuracy of all predictions. The size of the population doubles in each

FIGURE 1.8
Finite state machine of parity checker.

iteration as each parent gives rise to an offspring. The next generation is selected from top 50% of individuals. Now, these algorithms can handle numerical optimization problems as well.

1.7.4 Genetic Programming

Genetic programming was developed by Koza and suggests the self-evolution of a program. Instead of solving a problem from scratch, possible solutions, i.e., already present programs that fit the situation should be searched. Genetic programming aids this search.

As mentioned in Michalewicz et al. (1997), there are five major steps:
'Selection of terminals, selection of function, identification of evaluation function, selection of parameters, selection of terminal condition.'

Under evolution, programs are hierarchically structured, and the search space is a set of valid programs that can be represented as trees. A tree has functions and terminals as its parts with respect to the domain of problem; parts are chosen a priori in such a way that a portion of the created trees yields an answer. Initially, population consists of all the trees, and then a fitness value is assigned by analyzing performance of each tree based on predefined sets of test cases. The probability of selection increases with an increase in fitness value. Primarily, crossover is used to create offspring, and offspring are created by exchanging the subtrees of two parents. Other operators, such as mutation, permutation, and editing can also be applied. A set of procedures called automatically defined functions has been recently added for promoting reusability. They may perform complicated functions and can evolve at run time (Man et al., 1996).

1.7.5 Swarm Intelligence

A swarm is a large number of similar, simple agents communicating locally among themselves and their domain, with no central control to enable a worldwide intriguing conduct to rise.

As mentioned in Roli, Manfroni, Pinciroli, and Birattar (2011):
'Swarm intelligence can be defined as a relatively new branch of artificial intelligence that is used to model the collective behaviour of social swarms in nature such as ant colonies, honey bees, and bird flocks.'

Despite the fact that these agents are generally unsophisticated with constrained abilities alone, they communicate to each other with certain behavioural patterns to collaboratively accomplish fundamental tasks for their survival. The social collaboration among swarms can be either immediate or aberrant (Papahristou & Refanidis, 2011).

As mentioned in Wong (2015) 'Ant Colony Optimization (Dorigo & Gambardella, 1997), Particle Swarm Optimization (Poli, Kennedy, & Blackwell, 2007), and Bee Colony Optimization (Karaboga, Akay, & Ozturk, 2007), etc., are collectively known as Swarm Intelligence'. Unlike other evolutionary algorithms, selection is not involved, but a fixed population size is maintained across all generations. Individuals report their findings after each generation, which helps in setting the search strategy for the next generation. Initially, some algorithms were designed to find the shortest path, but later other applications were also developed (Wong, 2015).

1.8 Applications of Evolutionary Computing

Applications of evolutionary computing range from telecommunication networks to complicated systems, finance and economics, image analysis, evolutionary music, art parameter streamlining, and scheduling and logistics. Major applications are as follows.

1.8.1 Problems in Distributed and Connected Networks, Internetworks, and Intranetworks

For instance, in wavelength division multiplexing networks, fiber failure may result in huge data loss, which might be very important for secure applications. Thus, designing a virtual topology that can handle such cases is a challenge. Evolutionary algorithms may help in designing such a solution (Ergin et al., 2011).

1.8.2 Interaction with Complex Systems

An example is the design of Boolean networks for controlling behaviour of a robot. Networks may be designed by random local search techniques. Boolean networks can show rich and complex practices, regardless of their small depiction (Roli et al., 2011).

1.8.3 Finance and Economics

Macro-economic models depict the progression of financial amounts. The conjectures delivered by such models play a significant part for money-related and political choices. Genetic programming along with symbolic regression may be used to recognize variable interactions in vast datasets (Kronberger, Fink, Kommenda, & Affenzeller, 2011).

1.8.4 Computational Intelligence for Games

Backgammon is a board game, having a fixed number of checkers and a dice, but the rules for movement vary in different parts of the world. By using reinforcement learning and neural network function approximation, agents can be trained to learn evaluation functions of game position (Carlo, Politano, Prinetto, Savino, & Scionti, 2011).

1.8.5 Electronic Sign Automation

The continuous contracting of a semiconductor's hub makes their memory progressively inclined to electrical imperfections firmly identified with the internal memory structure. Investigating the impact of fabrication defects in future advancements and recognizing new classes of utilization fault models with their relating test successions is a time-expensive task and is mostly performed manually. A genetic algorithm may automate this procedure (Bocchi & Rogai, 2011).

1.8.6 Image Analysis and Signal Processing

Image analysis and signal processing is used, for example, to solve the problem of segmentation of lesions in ultrasound imaging (Bozkurt & Yüksel, 2015).

1.8.7 Art, Music, Architecture, and Design

For example, various sound synthesis parameters target different types of sounds. A genetic algorithm can optimize such parameters, which match the parameters of different sound synthesizer topologies (Castagna, Chiolerio, & Margaria, 2011).

Another example is 'a musical translation of the real morphology of the protein, that opens the challenge to bring musical harmony rules into the proteomic research field' (Gussmagg-Pfliegl, Tricoire, Doerner, Hart, & Irnich, 2011).

1.8.8 Search and Optimization in Transportation and Logistics

An example is a mail delivery routing problem. In this problem, all the routes are to be planned in advance for the postal customers. Routes should serve each customer exactly once with minimum cost. Another decision is of mode (car, bicycle, or moped) because everyone is not accessible by all modes. Thus, an algorithm may solve the route selection problem and the mode of transport to be selected (Ahmed & Glasgow, 2012).

1.9 Conclusion

It is very evident that nature teaches us many things, and we can draw inspiration from it. Evolutionary computation takes inspiration from biological evolution. Evolutionary computing is not the same old process; it follows a probabilistic approach and gains knowledge from each generation obtained on the basis of fitness value of individuals in a population. Less time and greater complexity of problems indicate the need for robust algorithms, which can be tailored according to the requirements. Evolutionary computing aims at providing solutions, which can solve complex problems in the least possible time. Solutions are not necessarily optimal. Despite a large variety of algorithms, 'survival of the fittest' remains a constant base for all of them. Recombination and mutation can be applied on selected candidates (on the basis of fitness) to obtain new candidates. There are several selection methods that can be applied. An algorithm starts with selecting a population and terminates when either the optimum is reached or any of the considerations such as CPU time, number of generations, etc., is fulfilled. There are many techniques and paradigms. Genetic algorithms help in providing multi-directional search, representing adaptation, and working on binary strings. Evolution strategies come with an interesting idea of float value vector pairs. Evolutionary programming helps in developing artificial intelligence to foresee the changes in a domain. The accuracy of prediction is measured with a pay-off function. Evolutionary programming represents individuals in the form of finite state machines. Genetic programming takes into account the set of possible solutions instead of solving problems from the scratch. The search space is represented in the form of trees, where each tree represents a program and a new tree is formed by applying operations on subtrees. Swarm intelligence maintains a fixed-size population across all generations. Knowledge from each generation helps in formulating new generations. Having so many techniques and paradigms offers a wide variety of applications in every field.

References

Ahmed, H., & Glasgow, J. (2012). *Swarm Intelligence: Concepts, Models and Applications (Technical Report)*. Kingston, Ontario, Canada: Queen's University.
Bäck, T. (1996). *Evolutionary Algorithms in Theory and Practice*. Oxford, UK: Oxford University Press.

Bäck, T., Fogel, D. B., & Michalewicz, Z. (2000). *Evolutionary Computation: Basic Algorithms and Operators*. Bristol, UK: Institute of Physics Publishing.

Banzhaf, W., Nordin, P., Keller, R. E., & Francone, F. D. (1998). *Genetic Programming: An Introduction*. San Francisco, CA: Morgan Kaufmann.

Bocchi, L., & Rogai, F. (2011). Segmentation of ultrasound breast images: Optimization of algorithm paramters. In P. Husbands & J.-A. Meyer (Eds.), *European Conference on the Applications of Evolutionary Computation* (Vol. 6624, pp. 163–172). Berlin, Germany: Springer.

Bozkurt, B., & Yüksel, K. A. (2015). *Parallel Evolutionary Optimization of Digital Sound Synthesis Parameters*. Berlin, Germany: Springer.

Carlo, S. D., Politano, G., Prinetto, P., Savino, A., & Scionti, A. (2011). Genetic defect based march test generation for SRAM. In C. Di Chio (Ed.), *Applications of Evolutionary Computation* (Vol. 6625, pp. 141–150). Berlin, Germany: Springer.

Castagna, R., Chiolerio, A., & Margaria, V. (2011). Music translation of tertiary protein structure: Auditory patterns of the protein folding. In EvoApplications 2011: EvoCOMNET, EvoFIN, EvoHOT, EvoMUSART, EvoSTIM, and Evo-TRANSLOG (Eds.), *European Conference on the Applications of Evolutionary Computation* (Vol. 6625, pp. 214–222). Berlin, Germany: Springer.

Darwin, C. (1859). *The Origin of Species*. London, England: John Murray

De Jong, K. A. (1975). An analysis of the behavior of a class of genetic adaptive systems (Doctoral dissertation, University of Michigan). *Dissertation Abstracts International*, 36(10), 5140B. (University Microfilms No. 76-9381)

Dorigo, M., & Gambardella, L. M. (1997). Ant colony system: a cooperative learning approach to the traveling salesman problem. *IEEE Transactions on Evolutionary Computation, 1*(1), 53–66.

Eberhart, R. C., Shi, Y., & Kennedy, J. (2001). *Swarm Intelligence* (The Morgan Kaufmann series in evolutionary computation). Elsevier.

Eiben, A. E., & Smith, J. E. (2007a). Chapter 1: Introduction. In A. E. Eiben & J. E. Smith (Eds.), *Introduction to Evolutionary Computing*. Berlin, Germany: Springer.

Eiben, A. E., & Smith, J. E. (2007b). Chapter 2: What is an evolutionary algorithm?. In A. E. Eiben & J. E. Smith (Eds.), *Introduction to Evolutionary Computing*. Berlin, Germany: Springer.

Ergin, F. C., Uyar, Ş, & Yayimli, A. (2011). Investigation of Hyper-heuristics for designing survivable virtual topologies in optical WDM networks. In EvoApplications 2011: EvoCOMNET, EvoFIN, EvoHOT, EvoMUSART, EvoSTIM, and EvoTRANSLOG (Eds.), *Applications of Evolutionary Computation*. Berlin, Germany: Springer.

Fogel, D. B. (1998). *Evolutionary Computation: The Fossil Record*. Piscataway, NJ: IEEE Press.

Gussmagg-Pfliegl, E., Tricoire, F., Doerner, K. F., Hartl, R. F., & Irnich, S. (2011, April). Heuristics for a real-world mail delivery problem. In *European Conference on the Applications of Evolutionary Computation* (pp. 481–490). Berlin, Heidelberg: Springer.

Holland, J. H. (1973). Genetic algorithms and the optimal allocation of trials. *SIAM Journal on Computing, 2*, 88–105.

Jones, G. (2002). *Genetic and Evolutionary Algorithms*. Wiley. doi:10.1002/0470845015.cga004

Koza, J. R. (1992). *Genetic Programming*. Cambridge, MA: MIT Press.

Karaboga, D., Akay, B., & Ozturk, C. (2007, August). Artificial bee colony (ABC) optimization algorithm for training feed-forward neural networks. In *International conference on modeling decisions for artificial intelligence* (pp. 318–329). Berlin, Heidelberg: Springer.

Kronberger, G., Fink, S., Kommenda, M., & Affenzeller, M. (2011). Macro-economic time series modeling and interaction networks. In C. Di Chio (Ed.), *Applications of Evolutionary Computation* (Vol. 6625, pp. 101–110). Berlin, Germany: Springer. doi:10.1007/978-3-642-20520-0_11

Man, K. F., Tang, K. S., & Kwong, S. (1996). Genetic algorithms: Concepts and applications. *IEEE Transactions on Industrial Electronics*, *43*, 519–534.

Michalewicz, Z., Hinterding, R., & Michalewicz, M. (1997) Fuzzy evolutionary computation. *Pedrycz [691]*, 3–31.

Papahristou, N., & Refanidis, I. (2011). Training neural networks to play backgammon variants using reinforcement learning. In *European Conference on the Applications of Evolutionary Computation* (Vol. 6624). Berlin, Germany: Springer.

Poli, R., Kennedy, J., & Blackwell, T. (2007). Particle swarm optimization. *Swarm intelligence*, *1*(1), 33–57.

Rechenberg, I. (1973). *Evolution Strategie: Optimierung Technisher Systeme Nach Prinzipien Des Biologischen Evolution*. Stuttgart, Germany: Frommann-Hollboog Verlag.

Roli, A., Manfroni, M., Pinciroli, C., & Birattar, M. (2011). On the design of Boolean network robots. In EvoApplications 2011: EvoCOMPLEX, EvoGAMES, EvoIASP, EvoINTELLIGENCE, EvoNUM, and EvoSTOC, *Applications of evolutionary computation*. Berlin, Germany: Springer. doi:10.1007/978-3-642-20525-5_5

Schwefel, H.-P. (1995). *Evolution and Optimum Seeking*. New York, NY: Wiley.

Wong, K. C. (2015). Evolutionary algorithms: Concepts, designs, and applications in bioinformatics: Evolutionary algorithms for bioinformatics. *arXiv preprint arXiv:1508.00468*.

Wright, S. (1932). *The roles of mutation, inbreeding, crossbreeding, and selection in evolution* (Vol. 1, No. 8). na: Blackwell Publishing.

2

Genetic Algorithms

Sandeep Kumar, Sanjay Jain, and Harish Sharma

CONTENTS

2.1 Overview of Genetic Algorithms

Nature-inspired computing is a combination of computing science and the knowledge from different academic streams, such as mathematics, biology, chemistry, physics, and engineering. This diversity inspires the researchers to develop innovative computational tools for hardware, software, or wetware and for the synthesis of patterns, behaviours, and organisms, to get rid of complex problems. The algorithms simulating processes in nature or inspired from some natural phenomenon are called nature-inspired algorithms. According to the source of inspiration, they are divided into different classes. Most of the nature-inspired algorithms are inspired by the working of natural biological systems, swarm intelligence, and physical and chemical phenomena. For that reason, they are called biology based, swarm intelligence based, or physics and chemistry based, according to their sources. Basically, their main source of inspiration is nature, and thus they are called nature inspired. More than a hundred algorithms inspired by nature are now available.

The process of natural selection is the imitation and differentiation of individuals due to continuous changes in phenotype. It is one of the fundamental mechanisms of evolution, the change in the genetic qualities, characteristic of a population over generations. The term 'natural selection' was popularized by Charles Darwin. In comparison to artificial selection, which is planned, natural selection occurs spontaneously. Natural selection is the process of selection of the best, and the discarding of the rest, or simply it is process of propagation of favorable characteristics through generations. Natural selection involves some major steps such as variation in traits, differential reproduction, and inheritence. Variation exists in all living individuals. It is an essential part of life. It occurs partially as mutations in the genome take place randomly due to the interaction of individuals with the environment, causing them to adapt to suitable changes. There are three primary sources of genetic variation: mutation, gene propagation, and sex. Mutation is basically changes in DNA, and even a small change in DNA can have a large effect. Generally, it takes place as a sequence of mutations. Mutation might lead to beneficial, neutral, or sometimes destructive changes because it never tries to supply required changes—it is a random process. Generally, a cell makes a copy of its DNA when it divides, and if the copy is not perfect, then there has been a mutation. Mutations may happen due to exposure to a particular chemical or to radiation as well. Another important source of genetic variation is gene movement from one population to another population. Gene combination in a population is introduced by sex. The behaviour of living organisms can also be molded by the process of natural selection. Bees' wiggle dance, the mating rituals of birds, and humans' ability to learn language are behaviours that have genetic mechanisms and are dependent on natural selection. Natural selection is also affected by human activities, and it forces the population to evolve and adapt, developing new features to survive. The possibility of survival or propagation into the next generation is determined by the fitness of individuals, and defines how good a certain genotype is. The fitness of an individual largely depends on his or her environment.

Genetic algorithms (GA) are a subcategory of evolutionary algorithm (EA) motivated by biological phenomena. Evolutionary algorithms follow the principle of 'survival of the fittest', which is a process of the natural selection of individuals from the population (Bäck & Schwefel, 1993). EAs have become very popular in the last two decades due to their straightforwardness and effortlessness employment. Modern meta-heuristics are very successful in the field of optimization especially for complex real-world problems. EAs are problem solving systems that use computational models of evolutionary procedures as fundamental components in their design and execution. Classifier systems, evolution strategies, evolutionary programming, genetic programming, and genetic algorithms are some very popular EAs nowadays. All these strategies have some similarity in working principles and different sources of inspiration. EAs start with a randomly initialized swarm. They then compute the fitness of all individuals and select the best individuals for the next iteration based on their fitness. This process is repeated until the termination criteria are satisfied. Almost 50 years of investigation and application have undoubtedly established that the search progression of natural advancement can produce highly robust, uninterrupted computer algorithms, although these models are simple generalizations of real biological processes. The backbone of evolutionary algorithms is the process of collective learning by individuals in a swarm. Each individual is analogous to a solution in the feasible search space for the considered problem. The initialization of a swarm is always random, and it progresses in the direction of best of the search space through randomized progressions of selection, recombination, and mutation. The quality information (fitness value) provided by the environment includes the search location and the selection of highly fitted solutions so that they can produce more offspring in comparison to low fitted solutions. The mechanism of recombination permits the intercourse of parental information even though sharing it with their offspring and mutation are known to improve the population.

'Evolutionary algorithm' is a widely used term to define computer-based problem solving strategies that use computational models of evolutionary processes as a crucial element in their design and application. First of all, Charles Darwin proposed the concept of EAs in 1859 inspired by the idea of natural selection and evolution. The process of evolution helps species to adapt to the environmental changes that continuously happen in the natural world. Climate, availability of food sources, the safety of the habitat, and the dangers of predators are the major environmental factors that influence the prospect for continued existence of a creature. Evolutionary algorithms mimic the basic model of this biological evolution process that happens in nature. To solve a specific problem, we produce an environment in which possible solutions can evolve. The environment is formed by the parameters of the problem and inspires the evolution of highly fitted solutions.

Major characteristics of EAs are:

- Flexible: Applicable to different types of problems.
- Robust: Can deal with noise and uncertainty.
- Adaptive: Can deal with dynamic environment.
- Autonomous: Occur without human intervention.
- Decentralized: Occur without a central control.

The most popular and successful evolutionary algorithm is the genetic algorithm (GA) (J. Holland, 1975). Additional wildly used evolutionary algorithms are evolution strategy (ES) (Rechenberg, 1965), evolutionary programming (EP) (Fogel, 2009), and genetic programming (GP) (Koza, 1990). GA is equally useful for continuous and discrete nature problems. GA is based on the biological phenomenon of gene propagation and the process of natural selection, and it tries to pretend the process of natural evolution at the genotype level, whereas ES and EP pretend the process of natural evolution occurs at the phenotype level. The differential evolution (DE) algorithm (Price, 1996) is one of the youngest evolutionary algorithms announced in recent times. DE is very efficient for continuous problems. In the elementary GA, selection operation is applied to the solutions appraised by the evaluation unit. The selection operator uses a greedy selection approach and selects the best one after comparing the fitness of parent and offspring. While, at the selection operation of the DE algorithm, each individual has an identical chance of being nominated as a parent because parents are selected randomly, i.e., the chance of selection is not subject to their fitness values. Application areas of EAs are very wide. Major application areas of EAs are:

- Planning routing, scheduling, packing
- Designing electronic circuits, neural networks, structure design
- Simulation modelling of economic interactions of competing firms in a market
- Identification of a function to medical data to predict future values
- Designing a controller for gas turbine engines, designing control systems for mobile robots
- Classification game playing, diagnosis of heart disease, Detecting Spam

Algorithm 1 and Figure 2.1 show the general scheme of an evolutionary algorithm in the form of algorithm and figure, respectively.

Algorithm 1: Evolutionary Algorithm
Step 1: Initialize population of individuals (arbitrarily generated first generation).

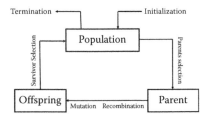

FIGURE 2.1
The general scheme of an evolutionary algorithm.

Step 2: Assess the fitness of each individual in current population.
Step 3: Repeat the following steps until the termination criteria meet
    The best fitted individuals (parent) are selected for reproduction.
    Generate new individuals (offspring) using crossover and muta-
    tion operations.
    Evaluate the fitness of offspring.
    Exchange the worst individuals with new ones.

The GA is one of the most successful evolutionary algorithms inspired by
natural selection. It was evolved in 1960 and proposed by John H. Holland
(1975) in 1970. It is used to get rid of multifaceted optimization problems
using bio-inspired operation such as mutation, crossover, and selection. Due
to the use of bio-inspired operations, it is also known as a bio-inspired swarm
intelligence algorithm. The concept of genetic algorithms is completely based
on Charles Darwin's quote 'It is not the strongest of the species that survives,

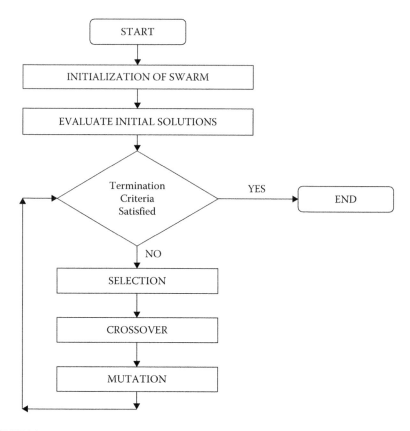

FIGURE 2.2
Flow chart for genetic algorithm.

nor the most intelligent, but the one most responsive to change.' The GA tries
to give the impression of the process of human evolution. The basic building
block of living things is cell. Every cell contains some set of chromosomes,
which are strings of DNA. These chromosomes denoted by binary strings.
During GA these binary strings are treated as solutions and go through
mutation, selection, and recombination for a number of iterations. The
Figure 2.2 shows the major steps followed by GA.

The GA mimics the survival of the fittest among individuals over
successive generations for solving a problem. Each generation of a GA
comprises a swarm of character strings that correspond to the chromosomes (that exist in living beings). Every individual characterizes a point in
the search space and a potential solution. Every individual should go
through a process of evolution in the population using the following steps:

- Individuals in the swarm compete for resources and mates.
- The successful individuals in each 'competition' get a chance to yield
  more offspring in comparison to poor performers.
- The best-fitted individuals spread their genes throughout the population in such a way that they generate better offspring.
- Therefore, each successive generation becomes better adapted to the
  environment.

2.2 Genetic Optimization

Optimization is the method of creating something that is as perfect as
possible. The best solution is selected from among all the available
alternatives. Optimization is the practice of discovering the smallest or
largest feasible solution that a given objective function can get within its
domain. The objective function that is to be optimized may have different
characteristics such as being linear, non-linear, differentiable, non-
differentiable, or fractional. Frequently, the function has to be optimized
in a recommended dominion that is indicated by several limitations in the
form of equalities or inequalities. This domain is termed the feasible
search space for the problem. Optimization is the process of deciding the
value of autonomous variables without violating constraints and providing an optimal solution as well. Therefore, the mathematical ideas for
searching for the optimum solution (maximum or minimum as per
requirement of objective function) of a function are called 'optimization
techniques'. Real-world problems with a large search space and multiple
constraints are not solvable without the help of robust and reliable
optimization techniques.

Optimization problems are very common in various areas, such as engineering design, communication, robotics, management, etc. The goal of optimization is to find the solutions that are optimal or nearest to optimal. Most complex optimization problems take more than one step to find a feasible solution. These steps follow some process that guides us through problem solving. Generally, this process is divided into diverse phases, which are accomplished one by one. Frequently used phases are identifying and describing objectives, building and solving models, and assessing and realizing solutions. Optimization problems arise in every field of engineering, science, and industry. Thus, it is necessary to develop some reliable, effective, and robust algorithms, which can get rid of these problems irrespective of their size and nature.

A single objective optimization problem may be formulated as follows:

$$Minimize \ (or \ maximize) f(x) : x = (x_1, x_2, ..., x_D)$$
$$Subject \ to \ x \in F, \ usually \ defined \ by$$
$$F = \{x \in \mathfrak{R}^D / h_i(x) = 0; \ and \ g_j(x) \geq 0\}$$
$$i = 1, 2, ..., n \ and \ j = m + 1, m + 2, ..., p$$
$$Where \ f, g_{m+1}, g_{m+2}, ..., g_p, h_1, h_2, ..., h_m \ are \ real \ valued$$
$$functions \ defined \ on \ \mathfrak{R}^D.$$

The GA is usually applied to maximization problems. There may exist zero or more optimal solutions for an optimization problem, i.e., sometimes there may not be any optimal solution or there may be multiple optimal solutions. A unique local optimal solution for a particular problem is also termed a global optimal solution. At least one solution will be a global solution if there are multiple optimal solutions for a problem. In the case of a linear programming problem (LPP), each and every local optimal solution is also the global optimal solution. While in non-linear programming problem (NLP), for a convex objective function (minimization) with a convex set of constraints, the global optima is the same as the local optima. Determining the global optimal solution of an NLP is more problematic than deciding the local optimal solution. Though, due to the real-world need, the exploration for the global optima is frequently required.

The genetic algorithm is one of the best tools for optimization. It is used for optimization in various fields, such as constrained (Morales & Quezada, 1998), global (Zhong, Liu, Xue, & Jiao, 2004), combinatorial (Han, Park, Lee, & Kim, 2001), and multi-objective optimization (Fonseca & Fleming, 1993).

2.2.1 Global Optimization

Global optimization methods concentrate on finding the best solution from all the local optima. It is a very tedious task to design a global optimization technique because there is no specific process or algorithm available for this designing process, and the criteria for optimum results are also not fixed. The literature contains a large number of heuristics and meta-heuristics to solve non-linear optimization problems. The approaches that presently exist in the literature for solving non-linear global optimization problems might be roughly categorized into two classes: deterministic and probabilistic methods. The deterministic strategies give us a guarantee of the global optimum. Probabilistic methods don't give us a guarantee of optimum, but they provide the nearest to the optimum solution. This is attained by supposing that the better solutions are in proximity to the solution search space, and this is true for most of the real-world problems. These probabilistic methods use random components for fluctuations and are also referred to as stochastic algorithms. A balanced approach between exploration of search space and exploitation of best feasible solutions found so far is considered to be most successful in this class.

The GA is most suitable for global optimization because it is essentially parallel, whereas almost all other probabilistic algorithms are serial and can explore only one direction at a time in the solution space. Although, GA can discover the solution space in many directions at the same time because it has multiple offspring. The GA is able to explore the feasible solution search space despite a failure in some direction. The GA is predominantly compatible with cracking problems with very large feasible solutions because it can assess many schemas simultaneously. A GA can handle a problem with a complex fitness landscape such as a fitness function that is noisy, is discontinuous, changes over time, or has multiple local optima. The most important feature of a GA is that, without knowing about problem, it can reach the nearest to optimum result.

The GA is a global optimization technique and adaptive strategy by nature. Tsai et al. (2004) conceived the hybrid Taguchi-genetic algorithm for global optimization. The proposed approach added a new step namely Taguchi between the mutation and crossover steps. The approach was tested over a global optimization benchmark problem with different dimensionality. Zhong et al. (2004) integrated GA with a multi-agent system to solve global optimization problems and get good results for high-dimension problems. Kim et al. (2007) hybridized GA with bacterial foraging to extend GA for global optimization. Leung and Wang (2001) proposed orthogonal GA with quantization for global numerical optimization.

2.2.2 Constrained Optimization

Variables every so often have restrictions or constraints. Constrained optimization combines variable equalities and inequalities into the cost function. A constrained variable frequently transforms into an unconstrained variable through a transformation of variables. Genetic algorithms are developed and most suitable for unconstrained optimization problems. It is a challenging task to apply GA to constrained optimization problems. Several strategies have been introduced in GA to handle constrained optimization such as the penalty-based method, searching feasible solutions, and preserving the feasibility of solutions. The constrained problem can be transformed into an unconstrained one using penalty function in a couple of ways Yeniay (2005). First method is additive:

$$eval(\bar{x}) = \begin{cases} f(\bar{x}), & if\ \bar{x} \in F \\ f(\bar{x}) + p(\bar{x}), & otherwise \end{cases} \tag{2.1}$$

Where $p(\bar{x})$ denotes penalty, in case of no violation it is zero else have some positive value. The second method of penalty function is multiplicative as shown in equation (2.2).

$$eval(\bar{x}) = \begin{cases} f(\bar{x}), & if\ \bar{x} \in F \\ f(\bar{x}) \times p(\bar{x}), & otherwise \end{cases} \tag{2.2}$$

In this case $p(\bar{x})$is one if no violation else it is greater than one. The multiplicative penalty function is more popular in comparison to additive penalty method among researchers. The death penalty method (Bäck, Hoffmeister, & Schwefel, 1991) is one popular method; it simply discards unachievable solutions from the population and prevents the occurrence of an unfeasible solution. This method is suitable for convex possible search space or a realistic part of the entire search space and is less fitted for highly constrained problems. A static penalty method was conceived by Homaifar et al. (1994), in which levels of violation demarcated by the user. It engenders L levels of violation for all constraint and the penalty coefficient (R_{ij}) for every level of violation and each constraint. Coefficients are assigned in proportion to the level of violation. Finally, it evaluates each individual by using equation (2.3).

$$eval(\bar{x}) = f(\bar{x}) + \sum\nolimits_{j=1}^{m} R_{ij}[0, g_j(\bar{x})]^2 \tag{2.3}$$

This method requires large number of parameter settings, for n constraints $n \times (2L + 1)$ parameter must be set.

A dynamic penalty function was proposed by Joines and Houck (1994) to assess individuals at each generation whose value depends on generation counter (t).

$$eval(\bar{x}) = f(\bar{x}) + (Ct)^\alpha SVC(\beta, \bar{x}) \tag{2.4}$$

In equation (2.4) the value of constants C, α, and β are decided by user. The function $SVC(\beta, \bar{x})$ is determined using some complex functions (Joines &

Houck, 1994). In this dynamic penalty method, the penalty increases with each generation. A varying fitness function in genetic algorithms was introduced by Kazarlis and Petridis (1998) that is dynamic in nature. The penalty function is defined in (Kazarlis & Petridis, 1998) with a severity factor (A), weight factor for the ith constraint (W_i), and a degree of violation d (S_i).

$$eval(\overline{x}) = f(\overline{x}) + V(g)\left[A\sum_{i=1}^{m}(\delta_i W_i \phi(d_i(S))) + B\right]\delta_s \qquad (2.5)$$

A new penalty method (GENECOP II), inspired by simulated annealing, was proposed by Michalewicz and Attia (1994). This method separates all the linear and non-linear equalities and inequalities. Then it randomly selects an initial point that satisfies the linear constraints and identifies a set of active constraints (A). The population evolved using equation (2.6).

$$eval(\overline{x}, \tau) = f(\overline{x}) + \frac{1}{2\tau}\sum_{j\in A}f_j^2(\overline{x}) \qquad (2.6)$$

Michalewicz and Attia (1994) differentiate among linear and non-linear constraints. Here in each iteration, only active constraints are considered. Selective pressure on unattainable solutions rises with diminution in temperature. The remarkable thing in this strategy is that no diversity of the initial generation that consists of multiple copies of a solution fulfilled all linear constraints. The best solution found so far is treated as the starting point for the next generation. The algorithm is terminated at a previously determined freezing temperature.

A penalty function with high penalties for unfeasible solutions has been proposed by Gen and Cheng (1996).

The proposed evaluation function is defined in equation (2.7).

$$eval(\overline{x}) = f(\overline{x})\left[1 - \frac{1}{q}\sum_{i=1}^{q}\left(\frac{\Delta b_i(\overline{x})}{\Delta b_i^{max}}\right)^k\right] \qquad (2.7)$$

Here, violation of ith constraint is denoted by $\Delta b_i(\overline{x})$ and maximum violation constraint is denoted by Δb_i^{max}. The approach proposed by Gen and Cheng (1996) is not dependent on the nature of the problem. It provides better diversity to the initial population and prevents infeasible solutions with high penalty. Segregated GA has been proposed by Le Riche et al. (1995) with two penalty parameters in different populations. It tries to maintain an average penalty and avoid extremely low and high penalties. This method is able to converge feasible and infeasible solutions simultaneously.

Practical optimization problems have constraints that must be fulfilled by the solution of the problem. Many researchers proposed different techniques for constraint handling, but each and every technique has pros and cons. The penalty function method is a widespread constraint handling method among users.

2.2.3 Combinatorial Optimization

Combinatorial optimization is the procedure of searching optima for an objective function with a discrete domain but a very large search space. Some popular combinatorial optimization problems are: the travelling salesman problem, bin-packing, integer linear programming, job-shop scheduling, and the Boolean satisfiability problem. Due to the large search space, it is not possible to solve these problems by exhaustive search methods. Anderson and Ferris (1994) applied GA to solve the assembly line balancing problem (ALBP) using both parallel and serial implementation. The ALBP is mostly used in the design of manufacturing lines with a large number of stations. In order to solve the ALBP, (Anderson & Ferris, 1994) GA went through some modifications, especially for worst fitness, and used a very high rate of mutation. Another important and complex combinatorial optimization problem is the travelling salesman problem (TSP). The TSP is one of the oldest and class NP-complete problem. It is very easy to define but very tough to crack, due to large search space, and it became a benchmark combinatorial optimization problem. Larranaga et al. (1999) reviewed various strategies to solve the TSP using GA and proposed some modifications in crossover and mutation operators. Binary and path representation were considered for TSP (Larranaga et al., 1999). Binary representation is not considered as good way of representation because it becomes out of control when handling large instances of problems and also sometimes it requires repairing. The second representation makes use of the location of cities and their relative order. Some other representations also used (Larranaga et al., 1999) include adjacency, ordinal, and matrix. A quantum-inspired GA was proposed by Talbi et al. (Talbi, Draa, & Batouche, 2004) to solve the TSP. This strategy combined the concept of quantum computing with GA. Katayama et al. (2000) introduced complete sub-tour exchange crossover in GA in order to reinforce local search efficiently as well as global search. Moon et al. (2002) used precedence constraints in TSP and applied GA as a topological search.

Another popular combinatorial optimization problem is minimum spanning tree problem. The process of finding a least-cost spanning tree in a weighted graph is known as the MST problem. This problem has been well deliberated, and several well-organized polynomial-time algorithms have been established by Prim, Kruskal, and Dijkstra (1999). Zhou and Gen (1999) proposed a variant of GA to get rid of MST with multiple criteria. Zhou et al. (1996) introduced a degree-constrained GA and applied it to solve MST with higher efficiency. A weighted coding was introduced in GA to find MST by

Raidl and Julstrom (2000). Other combinatorial optimization problems that are solved using GA are the assignment problem (Chu & Beasley, 1997), the vehicle routing problem (Baker, B. M. & Ayechew), the job-shop scheduling problem (Pezzella, Morganti, & Ciaschetti, 2008), the constraint satisfaction problem (Eiben, Raué, & Ruttkay, 1994), and many more.

2.2.4 Multi-Objective Optimization

Multi-objective optimization problems (MOOP) are problems in the field of optimization that involve two or more objective functions to be optimized all at once. These occur in various arenas of science as well as economics, engineering, and logistics, where optimal decisions are required to be taken in the presence of trade-offs between more than one contradictory objective. If one wants to optimize more than one (say K) objectives such that the objectives are contrasting, and the decision-maker has no perfect inclination of the objectives relative to each other. In many realistic problems, considered objectives clash with each other. Hence, optimizing an n-dimensional decision variable vector (say X) pertaining to a single objective often results in undesirable results with regard to the other objectives. Consequently, obtaining a perfect multi-objective solution that concurrently optimizes each objective function is more or less very difficult. A judicious solution to a multi-objective problem is to identify a solution set, each one of which satisfies the objectives at an adequate level without being subjugated by any other solution.

Recently Konak et al. (2006) and Tamaki et al. (1996) presented a state-of-the-art study of multi-objective optimization using GA. A rank-based fitness computation method was incorporated in GA to solve MOOP by Fonseca and Fleming (1993). In this method, it is assumed that best-fitted solutions have optimal solutions in their proximity. A non-dominated sorting based multi-objective GA (abbreviated as NSGA-II) has been proposed by Deb et al. (2000). The proposed approach is very fast and elitist and can deal with $O(MN^3)$ computational complexity very easily, with population size N and M objectives. Horn et al. (1994) anticipated a niched pareto GA that is very efficient while finding a pareto optimal set. Altiparmak et al. (2006) proposed a multi-objective approach to solve supply chain network issues for a plastic producer company. A modified GA for multi-objective optimization was conceived by Sardiñas et al. (2006) and used for the purpose of optimal cutting parameters, which are very important elements in any process planning for metal parts. The GA is one of the most popular heuristic methods to address multi-objective design and optimization problems. Jones et al. (2002) stated that more than 90% of the methodologies for multi-objective optimization are meant to fairly accurately represent the true Pareto front for the fundamental problem. A majority of these used a meta-heuristic technique, and 70% of all meta-heuristics methodologies were built on evolutionary approaches Altiparmak et al. (2006).

2.3 Derivation of Simple Genetic Algorithm

The GA is one of the popular optimization algorithms used to find the maximum or minimum of a function. The basic components of a GA are: population of chromosomes, fitness function and selection, and recombination and mutation operators.

Basic step in GA are: (Holland, 1975)

1. Identify the number of chromosomes, generation, mutation rate, and crossover rate values.
2. Generate chromosomes.
3. Repeat step 4 to step 7 until the termination criteria meet.
4. Evaluate the fitness values of the chromosomes by calculating the objective function.
5. Apply a selection operator.
6. Apply a crossover operator.
7. Apply a mutation operator.
8. Memorize the best chromosomes identified so far.

The complete process of GA can be depicted using an example. Consider an equality $p + 2q + 3r + 4s = 30$, find values of p, q, r, and s such that they fulfill the above equation. This equation can be formulated as an objective function $f(x) = p + 2q + 3r + 4s - 30$. In this case there are four variables; thus a chromosome can be constructed as follows:

| p | q | r | s |

The values of variables p, q, r, and s delimited in the range of 0 to 30 to speed up the calculation process. The first step of GA is initialization; let us assume that population size is 6. Then engender gene values (p, q, r, and s) for six chromosomes randomly.

$Chromosome_1 = [p; q; r; s] = [23;07;13;18]$
$Chromosome_2 = [p; q; r; s] = [04;21;08;07]$
$Chromosome_3 = [p; q; r; s] = [20;11;17;14]$
$Chromosome_4 = [p; q; r; s] = [01;19;18;09]$
$Chromosome_5 = [p; q; r; s] = [10;09;14;17]$
$Chromosome_6 = [p; q; r; s] = [08;09;22;12]$

After random initialization of chromosome, they are evaluated using their objective function.

f_obj_1 = ((23 + 2*07 + 3*13 + 4*18)– 30) = 118
f_obj_2 = ((04 + 2*21 + 3*08 + 4*07)– 30) = 68
f_obj_3 = ((20 + 2*11 + 3*17 + 4*14)– 30) = 119
f_obj_4 = ((01 + 2*19 + 3*18 + 4*09)– 30) = 99
f_obj_5 = ((10 + 2*09 + 3*14 + 4*17)– 30) = 108
f_obj_6 = ((08 + 2*15 + 3*23 + 4*12)– 30) = 125

After evaluation, the next step is selection of best fitted solution. The fitness of an individual is evaluated using its function value.

$$Fitness_i = \frac{1}{1 + f\_obj_i} \tag{2.8}$$

$Fitness_1$ = 1/(1+f_obj_1) = 1/(1+118) = 0.0084
$Fitness_2$ = 1/(1+f_obj_2) = 1/(1+68) = 0.0145
$Fitness_3$ = 1/(1+f_obj_3) = 1/(1+119) = 0.0083
$Fitness_4$ = 1/(1+f_obj_4) = 1/(1+99) = 0.0100
$Fitness_5$ = 1/(1+f_obj_5) = 1/(1+108) = 0.0092
$Fitness_6$ = 1/(1+f_obj_6) = 1/(1+125) = 0.0079

Total Fitness = 0.0084 + 0.0145 + 0.0083 + 0.0100 + 0.0092 + 0.0079 = 0.0583
   The probability of selection for an ith chromosome is computed using the following formula:

$$P_i = \frac{Fitness_i}{Total\ Fitness} \tag{2.9}$$

P_1 = 0.0084/0.0583 = 0.144
P_2 = 0.0145/0.0583 = 0.248
P_3 = 0.0083/0.0583 = 0.142
P_4 = 0.0100/0.0583 = 0.172
P_5 = 0.0092/0.0583 = 0.158
P_6 = 0.0079/0.0583 = 0.136

It can be observed from above fitness and probabilities that the second chromosome has the highest fitness and thus more chances of selection based on probability among the set of six. Now cumulative probability is required in order to apply roulette wheel selection approach.

$Cumulative\_Prob_1$ = 0.144
$Cumulative\_Prob_2$ = 0.144 + 0.248 = 0.392

$Cumulative\_Prob_3 = 0.392 + 0.142 = 0.534$
$Cumulative\_Prob_4 = 0.534 + 0.172 = 0.706$
$Cumulative\_Prob_5 = 0.706 + 0.158 = 0.864$
$Cumulative\_Prob_6 = 0.864 + 0.136 = 1.000$

Now a random uniform number (R) is generated for each chromosome. They are:

$R_1 = 0.453$
$R_2 = 0.024$
$R_3 = 0.901$
$R_4 = 0.142$
$R_5 = 0.008$
$R_6 = 0.768$

Now a new chromosome is selected by comparing R_i and $Cumulative\_Prob_j$ such that if R_i is greater than C_j and smaller than C_k, then $Chromosome_k$ is elevated in the next generation as a new chromosome in the new population.

$New\_Chromosome_1 = Chromosome_3$
$New\_Chromosome_2 = Chromosome_1$
$New\_Chromosome_3 = Chromosome_6$
$New\_Chromosome_4 = Chromosome_1$
$New\_Chromosome_5 = Chromosome_1$
$New\_Chromosome_6 = Chromosome_5$

Thus, new population of chromosomes is as follow:

$Chromosome_1 = [20;11;17;14]$
$Chromosome_2 = [23;07;13;18]$
$Chromosome_3 = [08;09;22;12]$
$Chromosome_4 = [23;07;13;18]$
$Chromosome_5 = [23;07;13;18]$
$Chromosome_6 = [10;09;14;17]$

The next step is crossover; it is controlled by a parameter 'crossover rate'. Assume that here the crossover rate is 30% (i.e., 0.3). A chromosome is selected for crossover if an arbitrarily generated number (in the range [0, 1]) is less than the crossover rate.
Assume that the following random number is generated.

$R_1 = 0.290$
$R_2 = 0.645$
$R_3 = 0.101$
$R_4 = 0.365$

$R_5 = 0.907$
$R_6 = 0.267$

Based on the above random numbers, chromosomes R_1, R_3, and R_6 are selected for the crossover operation. The crossover between the following pair will be performed.

$$Chromosome_1 \times Chromosome_3$$
$$Chromosome_3 \times Chromosome_6$$
$$Chromosome_6 \times Chromosome_1$$

Before performing the crossover, we need to identify the crossover point (in the range 1 to length of chromosome), here we use a one-cut point. Let the following crossover points be generated randomly:

$$C_1 = 2$$
$$C_2 = 1$$
$$C_3 = 1$$

$$
\begin{aligned}
Chromosome_1 \ &= \ Chromosome_1 \times Chromosome_3 \\
&= \ [20; \ 11; \ 17; \ 14] \ \times \ [08; \ 09; \ 22; \ 12] \\
&= \ [20; \ 11; \ 22; \ 12] \\
Chromosome_3 \ &= \ Chromosome_3 \times Chromosome_6 \\
&= \ [08; 09; 22; 12] \times [10; 09; 14; 17] \\
&= \ [08; 09; 14; 17] \\
Chromosome_6 \ &= \ Chromosome_6 \times Chromosome_1 \\
&= \ [10; 09; 14; 17] \times [20; 16; 11; 14] \\
&= \ [10; 11; 17; 14]
\end{aligned}
$$

After implementing crossover operation, the new population of chromosomes will be as follows:

$Chromosome_1 = [20;11;22;12]$
$Chromosome_2 = [23;07;13;18]$
$Chromosome_3 = [08;09;14;17]$
$Chromosome_4 = [23;07;13;18]$
$Chromosome_5 = [23;07;13;18]$
$Chromosome_6 = [10;11;17;14]$

The nest step of GA is mutation; a parameter mutation rate decides which individuals will be mutated.

Total genes = Number of genes in chromosomes * size of population
Number of mutations = Mutation rate * Total genes
Assume that in this case the mutation rate is 15% (i.e., 0.15).
Total genes = 4 * 6 = 24

$$Number\ of\ Mutations = 0.15 * 24 = 3.6 \approx 4$$

Suppose four randomly generated numbers are 4, 11, 13, and 21; thus, the chromosomes that will be mutated are replaced by an arbitrary number and generate new chromosomes as shown below:

$Chromosome_1 = [20;11;22;\textbf{05}]$
$Chromosome_2 = [23;07;13;18]$
$Chromosome_3 = [08;09;\textbf{11};17]$
$Chromosome_4 = [\textbf{03};07;13;18]$
$Chromosome_5 = [23;07;13;18]$
$Chromosome_6 = [\textbf{02};11;17;14]$

These new chromosomes will go through a progression similar to that of the preceding generation of chromosomes, including estimation, selection, crossover, and mutation, and at the end will yield a new generation of chromosomes for the next iteration. This process will be reiterated until the termination criteria meet. Assume that after 100 iterations, GA arrives at the best chromosomes [02; 06; 04; 01] for the considered example. The solution satisfies the given problem:

$p + 2q + 3r + 4s = 30$
$2 + (2 * 6) + (3 * 4) + (4 * 1) = 30$

Thus, GA is able to find a solution for a problem that satisfies the given equality.

2.4 Genetic Algorithms vs. Other Optimization Techniques

The GA is completely different from classical optimization techniques. It is global optimization strategy and adaptive in nature. The GA is considered a stochastic algorithm because it makes use of arbitrariness to generate initial solutions and new individuals, and to select individuals. Other nature-inspired algorithms are also stochastic in nature such as artificial bee colony, differential evolution, and ant colony algorithms. Most of the meta-heuristics use variables, whereas GA works with string coding of variables. Therefore, that coding discretizes the search space, while the function is continuous. It makes use of a pool of individual solutions, called the population, whereas simple random search techniques use a single solution. The good solution of the earlier iteration is accentuated using a reproduction operator and proliferated adaptively using crossover and mutation operators. It needs only objective function values to find an optimal solution. The evolution of the search process reduces the search space adaptively, which is a very important characteristic of GA.

Eberhart and Kennedy (1995) developed the PSO in the mid-1990s while trying to simulate the choreographed, elegant gesture of swarms of birds as a chunk of a socio-cognitive study exploring the concept of 'collective intelligence' in living populations. The GA was publicized by John Holland (1975) and his colleagues and students in the mid-1970s at the University of Michigan. The GA is stimulated by the principles of genetics and natural evolution and mimics the process of reproduction in nature, whereas the PSO is stimulated by the capability of schools of fish, flocks of birds, and herds of animals to search good food sources, acclimate to their environment, and search for safe territory by employing an 'information sharing' tactic, thus generating an evolutionary improvement. In contrast to the GA with its binary encoding, in PSO, the design variables can take any values, even sometimes outside their boundary constraints, which depend on their current location in the design space and the calculated velocity vector. If velocity vector grows very quickly, then design variables cross boundaries, which results in divergence. The performance of PSO heavily depends on initialization of a swarm because a diversified swarm may lead to better convergence speed, but it is not mentioned in literature what the size of the swarm should be. Generally, it is taken to be 10 to 50, but there is no firm recommendation. Similar to the PSO, the GA starts with an arbitrarily engendered population that evolves with each generation. In order to perform optimization, it evolves the population from one generation to another using the three operators.

R. Storn and K. Price (1996) proposed an evolutionary algorithm, namely differential evolution (DE) while solving the Chebyshev polynomial fitting problem. In DE, next-generation solutions evolve from the previous generation. Real-valued problems over a continuous search space can be easily cracked by DE in comparison to GA. In the case of DE, the difference of two randomly selected vectors with a scaling factor added in a third randomly selected vector. Similar to GA, the population is initialized randomly. The basic advantage of DE is that the parameters F and CR do not require the same fine-tuning that is essential for other evolutionary algorithms. Even though the idea of the GA is not excessively intricate, the individual parameters and employment of the GA typically entail a huge amount of fine-tuning. Another major difference between GA and DE is that GA is swarm based bio-inspired, but DE is not swarm based; however, it is a bio-inspired algorithm.

The GA is one of the oldest nature-inspired algorithms. After the development of GA, a number of nature-inspired algorithms were developed to solve complex optimization problems, but GA is still better than other algorithms for some problems.

2.5 Pros and Cons of Genetic Algorithms

The strongest point of GAs is that they can search in parallel. The GAs also involve some form of recombination, because this allows the conception of new solutions that have, by advantage of their parent's success, a higher possibility of resulting in a good performance. The crossover operator can be considered the main genetic operator, whereas mutation is less important. Crossover tries to retain the constructive characteristics of individual solutions and to eradicate unwanted components, although the randomness of mutation is undoubtedly more likely to destroy a strong candidate solution than to enhance it. Another advantage of GA is implied parallelism. The major advantages of the GA are:

- The GA is easy to understand.
- No derivatives required.
- The GA is basically parallel and easily implemented in a distributed environment.
- The GA can easily escape local minima.
- Always provides nearest to optimal.
- Applicable on a large class of problems, including real-world problems.

While it has numerous advantages, the GA has some drawbacks; some of them are listed here:

- The major drawback of GA is that it requires a suitable population according to the problem being considered.
- The GA requires a larger number of function evaluations than classical techniques.
- It needs a higher crossover rate (typically 80% to 95%).
- It has no guaranteed convergence despite a local minimum.
- It ought to discretize parameter space.
- A wise selection of mutation rate, selection function, and fitness function is required.

2.6 Hybrid Genetic Algorithms

The GA can cooperate with other NIAs; thus, it can be combined with them in order to improve performance. A hybrid algorithm might be used

as a local search strategy to improve exploration. Some popular hybridizations of GA with other NIAs are discussed here.

Kim et al. (2007) proposed a hybrid version of GA with a bacterial foraging
algorithm (BFA). The BFA was inspired by *Escherichia coli* bacteria's foraging
behaviour. The new GA-BF algorithm was used for tuning a PID controller of
an AVR system. The particle swarm optimization algorithm was hybridized
with GA (GA-PSO) and applied on multimodel optimization problems by
Kao and Zahara (2008). These algorithms work with the identical initial
population and randomly generate 4N individual solutions for an N-dimensional problem. These individuals might be considered chromosomes in the
case of GA, or particles in the case of PSO. In this approach, the first 2N
solutions are selected for real coded GAs and allowed to produce offspring
using crossover and mutation. These 2N offspring are adjusted in the remain-
ing 2N particles for implementation of PSO with the selection of the global
best individual. Juang (2004) designed persistent neural/fuzzy networks with
the help of a hybrid of GA and PSO. The PSO and GA were hybridized and
used to optimize a sketched crenelated horn antenna by Robinson et al.
(2002). A hybrid of GA with Tabu search as a local search was proposed by
Liaw (2000) and solved an open shop scheduling problem.

The GA's operators, such as mutation and crossover, are very popular in
other NIAs and are used to improve the performance of these algorithms.
Kumar et al. (2013) used the crossover operator in the ABC algorithm to
improve the exploration of the solution search space. The hill-climbing algorithm was hybridized with GA by Renders and Bersini (1994) in two different
ways for global optimization, using individual learning and modifying
genetic operators.

2.7 Possible Applications of Computer Science via Genetic Algorithms

The GA has a large list of applications. Its applications range from simple
computation to the designing of large complex circuits. Some applications of
GA are: automated designing, bioinformatics, climatology, clustering, gene
profiling, feature selection, game theory, scheduling, multi-dimensional, and
multimodal optimization, and networking and routing, among others.

The GA was used to solve assembly line balancing problem (ALBP) by
Anderson and Ferris (1994). A GA was also applied to solve one of the
complex problems, namely the travelling saleman problem (TSP), solved
by Larranaga et al. (1999). Talbi et al. (2004) proposed a quantum-inspired
GA for TSP and Moon et al. (2002) used precedence constraints in TSP and
applied GA as a topological search. The MST problem tackled by Zhou
et al. (1996, 1999) and by Raidl and Julstrom (2000) using GA. Some other
important applications of GA are: assignment problems (Chu & Beasley,

1997), vehicle routing problems (Baker & Ayechew, 2003), job-shop scheduling problems (Pezzella, Morganti, & Ciaschetti, 2008), and constraint satisfaction problems (Eiben, Raué, & Ruttkay, 1994).

The GA also has been used in designing of an obstinate neural network with the help of a hybrid version of GA and PSO by Juang (2004). Karr (1991) designed an adaptive fuzzy logic controller using GA. A side-lobe reduced in array-pattern synthesis using GA was performed by Yan and Lu (1997). Morris et al. (1998) proposed an automated docking application using a Lamarckian genetic algorithm and an empirical binding free energy function. An optimal sizing method for a stand-alone hybrid solar–wind system with LPSP technology was developed by Yang et al. (2008) using GA. An intrusion detection system was proposed by Li (2004) for network security using GA. The GA is very successful in the field of multi-objective optimization. Asadi et al. (2014) used it to develop a model for building retrofit. Gai et al. (2016) developed a cost-aware multimedia data allocation system for heterogeneous memory using GA in cloud computing. Ganguly and Samajpati (2015) implemented GA for the allocation of distributed generation on radial distribution networks under uncertainties of load and generation, using genetic algorithms. Elhoseny et al. (2015) used GA for balancing energy consumption in heterogeneous wireless sensor networks. Kavzoglu et al. (2015) used GA for selecting optimal conditioning factors in shallow translational landslide susceptibility mapping. Saha et al. (2015) improved test pattern generation for hardware Trojan detection using genetic algorithm and Boolean satisfiability. Faghihi et al. (2014) solved a complex optimization problem, namely construction scheduling, using a genetic algorithm established on a building information model. Thirugnanam et al. (2014) proposed a mathematical modelling of the Li-ion battery using a genetic algorithm approach for V2G applications. The GA has a very important role in drug designing and biochemistry; recently Nekoei et al. (2015) performed a QSAR study of VEGFR-2 inhibitors by using genetic algorithm-multiple linear regressions (GA-MLR) and a genetic algorithm-support vector machine (GA-SVM) with a state-of-the-art comparative study. Sharma et al. (2014) anticipated a survey on software testing techniques using GA.

The list of applications of GA is endless because it is widely applicable for real-world problems. Major areas of application of GA are: optimization, economics, neural networks, image processing, vehicle routing problems, scheduling, machine learning, and engineering applications.

2.8 Conclusion

The genetic algorithm is the most popular bio-inspired optimization technique that is stochastic in nature. Classical approaches take a very long time to search and optimize the solution that lies in most complex search spaces, even those applied in supercomputers. The GA is a vigorous search technique necessitating pint-sized knowledge to achieve search efficiently in huge search spaces or those where the reason is poorly understood. Specifically, the GA progresses through a population of points in contrast to the single point of focus of most search algorithms. Furthermore, it is valuable in the highly complicated region of non-linear problems. Its inherent parallelism (in evaluation functions, selections, and so on) permits the use of distributed processing machines. In GA, the selection of good parameter settings that work for a specific problem is very simple. It may be concluded that GA is the best choice for multi-objective optimization and applicable for a large class of problems.

References

Altiparmak, F., Gen, M., Lin, L. & Paksoy, T. (2006). A genetic algorithm approach for multi-objective optimization of supply chain networks. *Computers & Industrial Engineering, 51*(1), 196–215.

Anderson, E. J. & Ferris, M. C. (1994). Genetic algorithms for combinatorial optimization: The assemble line balancing problem. *ORSA Journal on Computing, 6*(2), 161–173.

Asadi, E., Da Silva, M. G., Antunes, C. H., Dias, L. & Glicksman, L. (2014). Multi-objective optimization for building retrofit: A model using genetic algorithm and artificial neural network and an application. *Energy and Buildings, 81*, 444–456.

Bäck, T., & Schwefel, H. P. (1993). An overview of evolutionary algorithms for parameter optimization. *Evolutionary Computation, 1*(1), 1–23.

Bäck, T., Hoffmeister, F., & Schwefel, H. P. (1991, July). A survey of evolution strategies. In *Proceedings of the Fourth International Conference on Genetic Algorithms* (Vol. 2, No. 9). San Mateo, CA: Morgan Kaufmann Publishers.

Baker, B. M., & Ayechew, M. A. (2003). A genetic algorithm for the vehicle routing problem. *Computers & Operations Research, 30*(5), 787–800.

Chu, P. C., & Beasley, J. E. (1997). A genetic algorithm for the generalized assignment problem. *Computers & Operations Research, 24*(1), 17–23.

Deb, K., Agrawal, S., Pratap, A., & Meyarivan, T. (2000, September). A fast-elitist non-dominated sorting genetic algorithm for multi-objective optimization: NSGA-II. In *International Conference on Parallel Problem Solving from Nature* (pp. 849–858). Berlin: Springer.

Eberhart, R., & Kennedy, J. (1995, October). A new optimizer using particle swarm theory. In *MHS'95., Proceedings of the Sixth International Symposium on Micro Machine and Human Science*, 1995. (pp. 39–43). IEEE.

Eiben, A. E., Raué, P. E., & Ruttkay, Z. (1994, June). Solving constraint satisfaction problems using genetic algorithms. In *IEEE World Congress on Computational Intelligence. Proceedings of the First IEEE Conference on Evolutionary Computation, 1994.* (pp. 542–547). IEEE.

Elhoseny, M., Yuan, X., Yu, Z., Mao, C., El-Minir, H. K., & Riad, A. M. (2015). Balancing energy consumption in heterogeneous wireless sensor networks using genetic algorithm. *IEEE Communications Letters, 19*(12), 2194–2197.

Faghihi, V., Reinschmidt, K. F., & Kang, J. H. (2014). Construction scheduling using genetic algorithm based on building information model. *Expert Systems with Applications, 41*(16), 7565–7578.

Fogel L. J., Owens A. J., Walsh M. J. (1966). *Artificial intelligence through simulated evolution.* New York: Wiley.

Fonseca, C. M., & Fleming, P. J. (1993, June). Genetic algorithms for multi-objective optimization: formulation discussion and generalization. *ICGA, 93,* 416–423.

Gai, K., Qiu, M., & Zhao, H. (2016). Cost-aware multimedia data allocation for heterogeneous memory using genetic algorithm in cloud computing. *IEEE Transactions on Cloud Computing.*

Ganguly, S., & Samajpati, D. (2015). Distributed generation allocation on radial distribution networks under uncertainties of load and generation using genetic algorithm. *IEEE Transactions on Sustainable Energy, 6*(3), 688–697.

Gen, M., & Cheng, R. A. Survey of penalty techniques in genetic algorithms, *Proceedings of the 1996 International Conference on Evolutionary Computation,* IEEE, 804–809, 1996.

Holland, J. H. (1975). Adaptation in natural and artificial systems: an introductory analysis with applications to biology, control, and artificial intelligence.

Homaifar, A., Qi, C. X., & Lai, S. H. (1994). Constrained optimization via genetic algorithms. *Simulation, 62*(4), 242–253.

Horn, J., Nafpliotis, N., & Goldberg, D. E. (1994, June). A niched Pareto genetic algorithm for multiobjective optimization. In *IEEE World Congress On Computational Intelligence. Proceedings of the First IEEE Conference on Evolutionary Computation, 1994.* (pp. 82–87). IEEE.

Joines, J. A., & Houck, C. R. (1994, June). On the use of non-stationary penalty functions to solve nonlinear constrained optimization problems with GA's. In *Proceedings of the First IEEE Conference on Evolutionary Computation, 1994. IEEE World Congress on Computational Intelligence.* (pp. 579–584). IEEE.

Jones, D. F., Mirrazavi, S. K., & Tamiz, M. (2002). Multiobjective meta-heuristics: An overview of the current state-of-the-art. *Eur J Oper Res, 137*(1), 1–9.

Juang, C. F. (2004). A hybrid of genetic algorithm and particle swarm optimization for recurrent network design. *IEEE Transactions on Systems, Man, and Cybernetics, Part B (Cybernetics), 34*(2), 997–1006.

Kao, Y. T., & Zahara, E. (2008). A hybrid genetic algorithm and particle swarm optimization for multimodal functions. *Applied Soft Computing, 8*(2), 849–857.

Karr, C. L. (1991). Design of an adaptive fuzzy logic controller using genetic algorithm. In *Proc. of the 4th Int. Conf. on Genetic Algorithm* (pp. 450–457).

Katayama, K., Sakamoto, H., & Narihisa, H. (2000). The efficiency of hybrid mutation genetic algorithm for the travelling salesman problem. *Mathematical and Computer Modelling, 31*(10–12), 197–203.

Kavzoglu, T., Sahin, E. K., & Colkesen, I. (2015). Selecting optimal conditioning factors in shallow translational landslide susceptibility mapping using genetic algorithm. *Engineering Geology, 192,* 101–112.

Kazarlis, S., & Petridis, V. (1998, September). Varying fitness functions in genetic algorithms: Studying the rate of increase of the dynamic penalty terms. In *International Conference on Parallel Problem Solving from Nature* (pp. 211–220). Berlin: Springer.

Kim, D. H., Abraham, A., & Cho, J. H. (2007). A hybrid genetic algorithm and bacterial foraging approach for global optimization. *Information Sciences*, 177(18), 3918–3937.

Konak, A., Coit, D. W., & Smith, A. E. (2006). Multi-objective optimization using genetic algorithms: A tutorial. *Reliability Engineering & System Safety, 91*(9), 992–1007.

Koza, J. R. (1990). *Genetic Programming: A Paradigm for Genetically Breeding Populations of Computer Programs to Solve Problems* (Vol. 34). Stanford, CA: Stanford University, Department of Computer Science.

Kumar, S., Kumar Sharma, V., & Kumari, R. (2013). A novel hybrid crossover-based artificial bee colony algorithm for optimization problem. *International Journal of Computer Applications, 82*(8), 18–25.

Larranaga, P., Kuijpers, C. M. H., Murga, R. H., Inza, I., & Dizdarevic, S. (1999). Genetic algorithms for the travelling salesman problem: A review of representations and operators. *Artificial Intelligence Review, 13*(2), 129–170.

Le Riche, R., Knopf-Lenior, C., & Haftka, R. T. A Segregated genetic algorithm for constrained structural optimization, *Proceedings of the Sixth International Conference on Genetic Algorithms*, Morgan Kaufmann, 558–565, 1995.

Leung, Y. W., & Wang, Y. (2001). An orthogonal genetic algorithm with quantization for global numerical optimization. *IEEE Transactions on Evolutionary Computation, 5*(1), 41–53.

Li, W. (2004). Using genetic algorithm for network intrusion detection. *Proceedings of the United States Department of Energy Cyber Security Group, 1*, 1–8.

Liaw, C. F. (2000). A hybrid genetic algorithm for the open shop scheduling problem. *European Journal of Operational Research, 124*(1), 28–42.

Michalewicz, Z., & Attia, N. (1994). Evolutionary optimization of constrained problems. In *Proceedings of the 3rd Annual Conference on Evolutionary Programming* (pp. 98–108). World Scientific.

Moon, C., Kim, J., Choi, G., & Seo, Y. (2002). An efficient genetic algorithm for the traveling salesman problem with precedence constraints. *European Journal of Operational Research, 140*(3), 606–617.

Morales, A. K., & Quezada, C. V. (1998, September). A universal eclectic genetic algorithm for constrained optimization. In *Proceedings of the 6th European Congress on Intelligent Techniques and Soft Computing* (Vol. 1, pp. 518–522).

Morris, G. M., Goodsell, D. S., Halliday, R. S., Huey, R., Hart, W. E., Belew, R. K., & Olson, A. J. (1998). Automated docking using a Lamarckian genetic algorithm and an empirical binding free energy function. *Journal of Computational Chemistry, 19*(14), 1639–1662.

Nekoei, M., Mohammadhosseini, M., & Pourbasheer, E. (2015). QSAR study of VEGFR-2 inhibitors by using genetic algorithm-multiple linear regressions (GA-MLR) and genetic algorithm-support vector machine (GA-SVM): A comparative approach. *Medicinal Chemistry Research, 24*(7), 3037–3046.

Pezzella, F., Morganti, G., & Ciaschetti, G. (2008). A genetic algorithm for the flexible job-shop scheduling problem. *Computers & Operations Research, 35*(10), 3202–3212.

Price, K. Differential evolution: A fast and simple numerical optimizer," In *Fuzzy Information Processing Society, 1996. NAFIPS. 1996 Biennial Conference of the North American*. IEEE, 1996, pp. 524–527.

Raidl, G. R., & Julstrom, B. A. (2000, March). A weighted coding in a genetic algorithm for the degree-constrained minimum spanning tree problem. In *Proceedings of the 2000 ACM symposium on Applied Computing* (Vol.1, pp. 440–445). ACM.

Rechenberg, I. (1965). Cybernetic solution path of an experimental problem. Royal aircraft establishment, library translation 1122. Reprinted in evolutionary computation the fossil record, 1998. Fogel, D ed. chap. 8, pp. 297–309.

Renders, J. M. & Bersini, H. (1994, June). Hybridizing genetic algorithms with hill-climbing methods for global optimization: Two possible ways. In *IEEE World Congress on Computational Intelligence. Proceedings of the First IEEE Conference on Evolutionary Computation*, 1994. (pp. 312–317). IEEE.

Robinson, J., Sinton, S., & Rahmat-Samii, Y. (2002). Particle swarm, genetic algorithm, and their hybrids: Optimization of a profiled corrugated horn antenna. In *IEEE Antennas and Propagation Society International Symposium*, 2002. (Vol. 1, pp. 314–317). IEEE.

Saha, S., Chakraborty, R. S., Nuthakki, S. S., & Mukhopadhyay, D. (2015, September). Improved test pattern generation for hardware trojan detection using genetic algorithm and boolean satisfiability. In *International Workshop on Cryptographic Hardware and Embedded Systems* (pp. 577–596). Berlin: Springer.

Sardinas, R. Q., Santana, M. R., & Brindis, E. A. (2006). Genetic algorithm-based multi-objective optimization of cutting parameters in turning processes. *Engineering Applications of Artificial Intelligence*, 19(2), 127–133.

Sharma, C., Sabharwal, S., & Sibal, R. (2014). A survey on software testing techniques using genetic algorithm. *International Journal of Computer Science Issues*, 10(1), 381–393, 2013.

Talbi, H., Draa, A., & Batouche, M. (2004, December). A new quantum-inspired genetic algorithm for solving the travelling salesman problem. In *IEEE ICIT'04. 2004 IEEE International Conference on Industrial Technology, 2004.* (Vol. 3, pp. 1192–1197). IEEE.

Tamaki, H., Kita, H., & Kobayashi, S. (1996, May). Multi-objective optimization by genetic algorithms: A review. In *Proceedings of IEEE International Conference on Evolutionary Computation*, 1996., (pp. 517–522). IEEE.

Thirugnanam, K., Tp, E. R. J., Singh, M., & Kumar, P. (2014). Mathematical modeling of Li-ion battery using genetic algorithm approach for V2G applications. *IEEE Transactions on Energy Conversion*, 29(2), 332–343.

Tsai, J. T., Liu, T. K., & Chou, J. H. (2004). Hybrid Taguchi-genetic algorithm for global numerical optimization. *IEEE Transactions on Evolutionary Computation*, 8(4), 365–377.

Yan, K. K., & Lu, Y. (1997). Sidelobe reduction in array-pattern synthesis using genetic algorithm. *IEEE Transactions on Antennas and Propagation*, 45(7), 1117–1122.

Yang, H., Zhou, W., Lu, L. & Fang, Z. (2008). Optimal sizing method for stand-alone hybrid solar–Wind system with LPSP technology by using genetic algorithm. *Solar Energy*, 82(4), 354–367.

Yeniay, Ö. (2005). Penalty function methods for constrained optimization with genetic algorithms. *Mathematical and Computational Applications*, 10 (1), 45–56.

Zhou, G., & Gen, M. (1999). Genetic algorithm approach on multi-criteria minimum spanning tree problem. *European Journal of Operational Research*, 114(1), 141–152.

Zhou, G., Gen, M., & Wu, T. (1996, October). A new approach to the degree-constrained minimum spanning tree problem using genetic algorithm. In *IEEE International Conference on Systems, Man, and Cybernetics*, 1996. (Vol. 4, pp. 2683–2688). IEEE.

Zhong, W., Liu, J., Xue, M., & Jiao, L. (2004). A multiagent genetic algorithm for global numerical optimization. *IEEE Transactions on Systems, Man, and Cybernetics, Part B (Cybernetics)*, 34(2), 1128–1141.

3

Introduction to Swarm Intelligence

Anand Nayyar and Nhu Gia Nguyen

CONTENTS

3.1 Biological Foundations of Swarm Intelligence

Swarm intelligence has attracted lots of interest from researchers in the last two decades. Many swarm intelligence-based algorithms have been introduced to various areas of computer science for optimization, gaining lots of popularity. The reasons behind the success of swarm intelligence–based algorithms are their dynamic and flexible ability and that they are highly efficient in solving nonlinear problems in the real world. Bonabeau, Dorigo, & Theraulaz (1999) defined swarm intelligence as *"The Emergent Collective Intelligence of groups of simple agents"*. Swarm intelligence is regarded as a study of intelligent computational systems derived from "collective intelligence". Collective Intelligence (CI) is defined as the group intelligence that emerges from the cooperation or collaboration of a large number of swarm agents, acting out the same protocol within the same environment. Examples of homogeneous swarm agents are ants, fish, termites, cockroaches, bats, and more. The two fundamental aspects, which are necessary properties of swarm intelligence are: self-organization and division of labour. Self-organization is regarded as a significant property of natural systems; it is organization that takes place without any requirement of central coordinating authority to perform required tasks. Swarm agents exploit self-organization to attain the best work coordination, the agility to perform work at high speeds and, above all, fault tolerance. Self Organization is based on four fundamental elements: positive feedback, negative feedback, fluctuations, and multiple interactions as defined by Bonabeau et al. (1999). Positive feedback (amplification) is basically defined as the promotion of smart solutions by allocating more agents to the same work. Negative feedback (counterbalance) enables the swarm agents to avoid the same behaviour or involvement in the same state. Fluctuations are highly useful towards randomness, errors and random walks. Multiple interactions (sharing information) is the condition of sharing the information among all other agents of search area. There exists some continuous agitation between positive and negative feedback, and it usually happens in various self-organizing conditions such as market analysis, the study of complex networks, neural networks, etc. The second property of swarm intelligence is termed division of labour, which is regarded as a parallel execution of various tasks by swarm agents.

In general terms, swarm intelligence, which is regarded as sub-area of artificial intelligence (AI), is primarily concerned with a design and an implementation of intelligent multi-agent systems following the footsteps of behaviour of natural social insects and animals like ants, termites, bees, bats, cockroaches, fish, and more. Every single member of the colony is termed as unsophisticated, but they tend to accomplish complex tasks by cooperation with other swarm agents. Example: Ants roam in arbitrary fashion manner in an open environment searching for food, and after locating a food source, return to their nest to communicate with other members of the colony regarding the food source. Members of other colonies make a suitable and optimized path by laying pheromone trail from their nest to the food source. Termites and wasps coordinate with each other to build sophisticated nests. The same is true for the ability of bees to orient themselves in an environment by generating awareness among other honey bees regarding a food source via the waggle dance.

The term swarm intelligence was first coined by Beni (1988) with regard to cellular robotic systems, where simple agents utilize the principle of self-organization via the nearest-neighbour interaction.

Swarm intelligence has led to the development of various models. From the point of view of computational intelligence, swarm intelligence models are computing models that have formulated algorithms for designing optimized solutions for distributed optimization problems. The primary objective of swarm intelligence is to design probability-based search algorithms.

All the swarm agents working collectively have a number of general properties Bai and Zhao (2006).

- Communication among swarm agents is highly indirect and of short duration.
- Agents communicate in a distributed manner without any central coordinating authority.
- There is a high degree of robustness among agents working in an orderly fashion.

Swarm intelligence models and methods are highly successful, especially in the area of optimization, which is suitable for various applications of computer science, industrial, and other scientific applications. Examples: student time table management systems, train transportation scheduling, the design of efficient telecommunication networks, computational biology, neural networks, artificial intelligence, machine learning, and deep learning problems.

The most basic optimization problem, which has led to the easiness of various test cases is the travelling salesman problem (TSP) Lenstra, Kan, and Shmoys (1985). The problem is simple: a salesman has to traverse a large number of cities, ensuring that the total distance between all traversed cities is minimal. It has been applied to various problems of biology, chemistry, and physics, and results were successful.

In a general methodology, any optimization problem P can be defined as integration of three components: S, Ω, f, where:

S: A search space defined over set of decision variables. These variables can deal with discrete or continuous optimization problems.

Ω: The set of constraints on the variables.

f: S $\rightarrow$ IR$^+$ is the objective function used to assign a positive value to search element of S.

The main objective of this chapter is to start with a brief discussion of the biological foundations and the concept of swarm intelligence. After that, various concepts revolving around swarm intelligence such as metaheuristics, collective intelligence of animals, and concepts revolving around swarm-self organization will be discussed. The chapter will embark a comprehensive coverage to the issues concerning swarm intelligence as well as concluding with a future direction of swarm intelligence by applying swarm intelligence to robotics i.e. Swarm Robotics.

3.2 Metaheuristics

3.2.1 The History of Metaheuristics

Metaheuristics is a somewhat under-appreciated but widely used and adapted term. It is also regarded as a "search" or "optimization" algorithm. The true "metaheuristic" is regarded as a specific design pattern which instills knowledge and is applied to develop various optimization algorithms. Metaheuristics is a sub-area of "stochastic optimization". Stochastic optimization is a combination of algorithms and techniques deploying the highest degree of randomness to find optimal solutions to hard problems.

It is highly important to understand the history of metaheuristics (Sörensen, Sevaux, & Glover, 2018). Since the 1940s, many different metaheuristics have been proposed, and a few are in development today.

Different problem-solving techniques have evolved since the inception of human beings, but using metaheuristics is regarded as scientific approach to solve the problems. The first technique under metaheuristics was designed and developed by Igno Rechenberg, Hans-Pail Schwefel and their colleagues from Germany called it "evolutionary strategy" Osman and Laporte (1996). The main objective of this technique is to find a solution to various optimization problems via computers. In the 1960s, Lawrence Fogel and his colleagues proposed "evolutionary programming" to use simulated evolution as learning process.

In the 1970s, the genetic algorithm was proposed in a book titled *Adaption in Natural and Artificial Systems* (Holland, 1992). Later in 1970s, Fred Glover (1977) proposed the scatter search method, which created a significant foundation for the design and development of various methods derived from recourse to randomization. These developments are called "evolutionary algorithms or computation".

The decades of the 1980s and 1990s saw a rapid development of various metaheuristic algorithms. Fred Glover, in 1980, proposed a metaheuristic algorithm called "tabu search" (Talbi, 2009), "simulated annealing" was proposed by S. Kirkpatrick, Gelatt, and Vecchi (1983). The artificial immune system (AIS) technique was developed by Farmer, Packard, and Perelson (1986). The major breakthrough came in the area of metaheuristics in 1992: Marco Dorigo proposed the ant colony optimization (ACO), based on a nature-inspired technique for optimization of problems, in his PhD thesis (1992). The technique is based on ants using a pheromone-based chemical trail to find the optimal path from their nest to a food source. In 1995, James Kennedy and Russel Eberhart proposed "particle swarm optimization (PSO)". It is based on the movement of organisms such as a flock of birds or a school of fish. The PSO algorithm works on the basis of a population called "swarm agents" within candidate solutions, known as particles. In 1994 in a book titled *Genetic Programming* (Koza, 1994), John Koza laid the foundations of the optimization technique called "machine learning," which is highly utilized today in various branches of computer science. In 1997, R. Storn proposed "differential evolution" (Storn & Price, 1997), which is regarded as the best optimization technique in various applications, compared to genetic algorithms.

Various significant improvements have been seen from 2000 onward by various researchers proposing various metaheuristic-based algorithms. The following is the list of proposed algorithms.

Significant improvements have been observed since 1983 onwards. The following Table 3.1 enlists metaheuristic-based and swarm intelligence-based algorithms proposed by researchers since year 1983 to 2017.

3.2.2 Introduction to Metaheuristics

The word "metaheuristics" is combination of two words "meta" and "heuristics". The Greek word "meta" means "beyond" or "upper level". The word "heuristics" is derived from Greek word "heuriskein", which means "to find" or "to discover". The term metaheuristics can be derived as "higher level discovery" which can perform better than "simple heuristics/discovery". The word "metaheuristics" was coined by Dr. Fred Glover in his paper "Heuristics for integer programming using surrogate constraints" in *Decision Sciences* (Glover, 1977). In this article, he proposed "metaheuristics" as a powerful strategy which lays the foundation for other heuristics to produce optimal solutions as compared to normal solutions. Before his research, algorithms having stochastic components were called heuristics, but these days, they are termed "metaheuristics".

To date, the most clear and comprehensive definition of metaheuristics was proposed by Sörensen and Glover (2013), as:

> A metaheuristic is a high-level problem-independent algorithmic framework that provides a set of guidelines or strategies to develop heuristic optimization algorithms. The term is also used to refer to a problem-specific implementation of a heuristic optimization algorithm according to the guidelines expressed in such a framework.

The word "metaheuristic" is utilized for two different prupuses. One is a high-level framework in which varied concepts and methodologies are combined and lay a strong foundation for the development of various

TABLE 3.1

Metaheuristics & Swarm Intelligence based algorithms

Year	Name of Algorithm	Proposed by
1983	Simulated annealing	Kirkpatrick et al.
1986	Tabu search	Glover
1989	Memetic algorithms	Moscato
1992	Ant colony optimization (ACO)	Dorigo
1995	Particle swarm optimization (PSO)	Kennedy & Eberhart
2000	Bacteria foraging algorithm	Kevin Passino
2001	Harmony search	Geem, Kim & Loganathan
2005	Artificial bee colony algorithm	Karaboga
	Glowworm swarm optimization	Krishnanand & Ghose
	Bees algorithm	Pham
2006	Shuffled frog leaping algorithm	Eusuff, Lansey & Pasha
2007	Imperialist competitive algorithm	Atashpaz-Gargari & Lucas
	River formation dynamics	Rabanal, Rodriguez & Rubio
	Intelligent water drops algorithm	Shah-Hoseeini
	Monkey Search	Mucherino & Seref
2009	Gravitational search algorithm	Rashedi, Nezamabadi-pour & Saryazdi
	Cuckoo search	Yang & Deb
	Group search optimizer	He, Wu & Saunders
2010	Bat Algorithm	Yang
2011	Spiral optimization algorithm (SPO)	Tamura & Yasuda
	Galaxy based search algorithm	Hameed Shah-Hosseini
2012	Flower pollination algorithm	Yang
	Differential search algorithm (DS)	Civicioglu
2013	Artificial cooperative search algorithm	Civicioglu
	Cuttlefish optimization algorithm	Eesa, Mohsin, Brifcani & Orman
2014	Artificial swarm intelligence	Rosenberg
2016	Duelist Algorithm	Biyanto
	Killer whale algorithm	Biyanto
2017	Rain water algorithm	Biyanto
	Mass and Energy balances algorithm	Biyanto
	Hydrological cycle algorithm	Wedyan et al.

novel optimization algorithms. The second denotes specific implementation of algorithm proposed on a framework to find solution to a specific problem.

According to Osman & Laporte (1996),

> Metaheuristics is formally defined as an iterative generation process that guides a subordinate heuristic by combining intelligently different concepts for exploring and exploiting the search space; different learning strategies are used to structure the information in order find efficient near-optimal solutions.

In the last few decades, there has been significant development and innovation of approximation solution methods – heuristics and metaheuristics.

Various algorithms have been proposed on the basis of metaheuristics such as: simulated annealing (SA), tabu search (TS), the memetic algorithm (MA), the artificial immune system (AIS), ant colony optimization (ACO), genetic algorithms (GA), particle swarm optimization (PSO), and differential evolution (DE), which are utilized by researchers for optimizing and finding solutions to various problems of computer science. A metaheuristic algorithm is regarded as an iterative process that lays paths and innovates operations of sub-heuristics to define efficient solutions. It manipulates a complete or incomplete single solution or set of solutions at each iteration. Today, the scope of metaheuristics has widened, and various methods have been proposed such as: ant colony systems, greedy randomized adaptive search, cuckoo search, genetic algorithms, differential evolution, bee algorithms, particle swarm optimization (PSO), etc.

3.2.3 Characteristics of Metaheuristics

In the olden days, problem solving was done via heuristic or metaheuristic approaches – by trial-and-error methodology. Many important discoveries, even those still utilized today, are the result of "thinking outside of the box", and sometimes merely arrived at by a simple accident; this approach is called heuristics. In our day-to-day life, we handle almost every problem via heuristics.

But now, heuristics are being taken to the next level, i.e. metaheuristics. The major reason behind the popularity of metaheuristics is that all the algorithms are designed and developed by considering nature-based systems, i.e., swarm agents, which include biological systems, and physical and chemical processes. The most important component of metaheuristics is that all the algorithms imitate the best features of nature.

The two major components of all metaheuristic algorithms are: intensification and diversification, or exploration and exploitation. The intensification, or exploration, phase revolves around searching the best possible solutions available and selecting the best solution among the available options. The diversification, or exploitation, phase ensures that the search space is more efficiently explored. The overall efficiency of the algorithm depends on the balance of these two components. If there is a little exploration, the system will have limited boundaries and it will be difficult to locate the global optimum. If there is too much exploration, the system will find issues in reaching convergence, and the performance of search, in the overall algorithm will be reduced. The basic principle of best search is "only the best survives", so a proper mechanism is required to achieve the best algorithm. The mechanisms should be run repeatedly in order to test the existing solution and locate new options to verify whether or not they are the best solutions.

The primary foundations of metaheuristic algorithms are "nature based" and apply either a population or a single solution to explore the search space. Methods using single solution utilize local search-based metaheuristics such as tabu search and simulated annealing, which share the property of describing a state in the search space during the search process. On the contrary, population-based metaheuristics, such as genetic algorithms, explore the search space through the evolution of a set of solutions in the search space.

The following are some of fundamental properties which characterize metaheuristics in highly simple manner:

- Metaheuristics lay the stepping stone that "guides" the entire process of search.
- The ultimate goal is to explore the search space efficiently in order to determine the optimal solutions.
- Metaheuristic algorithms are approximate and non-deterministic in nature.
- Techniques integrating metaheuristic algorithms range from simple local search procedures to high-end complex learning processes.
- Metaheuristics are not problem-specific and avoid getting trapped in confined areas of search space.
- Metaheuristic algorithms make use of domain-specific knowledge via heuristics and are controlled by upper-level strategy, and today's metaheuristics make use of search experience to achieve optimal solutions.

3.2.4 Classification of Metaheuristics

A metaheuristic algorithm can provide a highly optimized and efficient solution to any optimization problem, but only if there exists a proper balance between the exploitation of the search experience and the search space exploration to identify the areas with high-quality solutions with regard to problems. The only difference between various metaheuristic techniques is how they provide the balance.

According to Talbi (2009), the following are different ways to classify metaheuristic algorithms:

- Nature-inspired versus non-nature-inspired algorithms

The most active field of research in the area of metaheuristics is design and development of nature-based algorithms. The most recent proposed meta-heuristics is "evolutionary computation-based algorithms", which are based on natural systems. Examples include: simulated annealing, ant colony optimization (ACO), particle swarm optimization (PSO), and evolutionary algorithms.

Non-nature-inspired metaheuristic algorithms include: tabu search (TS) and iterated local search (ILS).

- Population-based versus single-point search

Another important basis that is used for classification of metaheuristic algorithms is the total number of solutions used at the same time. Algorithms working on single solutions are termed as "trajectory methods" and include local search-based metaheuristics. Examples: tabu search, variable neighbourhood search, and iterated local search.

Population-based methods are highly focused on the exploration of the search space. Examples: Genetic algorithms and particle swarm optimization (PSO).

- Dynamic versus static objective function

Metaheuristics can also be classified on the basis of the methodology of which they make use; dynamic versus static objective function. Algorithms with a static objective function such as PSO keep the objective function in the problem representation.

Algorithms with dynamic objective function like guided local search can modify the function during the search process.

- One versus multiple neighbourhood structures

Most of the metaheuristic algorithms work on one single neighbourhood structure and don't change their fitness landscape topology during search. Example: iterative local search.

Some metaheuristics make use of set of neighbourhood structures which can change the fitness landscape topology during search. Example: variable neighbourhood Search (VNS).

- Iterative versus greedy

Iterative metaheuristic algorithms are those that start from a particular set of solutions and keep on manipulating as the search process continues, such as particle swarm optimization (PSO) and simulated annealing (SA).

Greedy metaheuristic algorithms are those that start from empty solution, and at every stage a decision variable is assigned a value and is added to the solution set. Example: ant colony optimization.

- Memory Usage versus memory-less

The most significant parameter used to classify metaheuristic algorithms is memory usage or memory-less. Metaheuristic algorithms make use of search history, and it is important to know, whether they make use of memory or not.

Memory-less algorithms are those that make use of Markov process to determine the next course of action in the current state. Example: simulated annealing.

Memory-usage algorithms are those that keep track of all the moves and decisions undertaken, and they make use of memory. Nowadays, all the metaheuristic algorithms make use of memory, and they are termed powerful algorithms. Example: tabu search.

3.3 Concept of Swarm

The concept of swarm is particularly a biological one and its deep roots can be found in ethology (the study of animal behaviour). Which has been engaged in since the nineteenth century. Ethology is a mixture of diverse fields, including zoology (classification), anatomy (structure), and the study of ecosystems (context). Ethology brings together these three areas in order to understand how the living beings exist and interact with one another. Ethology involves not only studying specific animals, but also interactions within groups and between individuals, and also how they live with regard to environmental constraints.

Ethology is primarily concerned with a study of natural living organisms such as ants, bees, fireflies, bats, and many more. The study of social insects is somewhat complicated, but they make "intelligent social interactions cum structures" by performing division of labor, hierarchy of control, and task coordination.

In ant colonies and bee colonies, there is ant queen or queen bee, respectively, as well as the proper top-down management of task allocation and all sorts of social functions, such as: army ants, worker ants, search ants; and likewise in the case of bees, onlooker bees, worker bees, etc. In ant colonies, various functions are performed like- foraging, pheromone split, optimal path trail. Other intelligent swarms also perform varied functions like Nest building by wasps, coordinated flashing by fireflies and waggle dance in case of honey bees.

In the same manner, the study of bird flocking is another example of how efficient organization happens without any sort of centralized control. It is a combination of the ethology and computer science which leads to the study of behaviour of birds and finding suitable applications for proposing solutions to complex problems.

Swarm (Glover, 1977; Lévy, 1997) is generally regarded as group of similar animals having qualities of self-organization and decentralized control.

According to Bonabeau et al. (1999) the following are typical characteristics of a swarm:

- A swarm consists of many individuals working collectively and has the characteristic of self-organization.
- Individuals of swarm are homogeneous (identical) such as a robotic swarm or some species of same animals.
- Individuals of the swarm, when compared with the abilities of whole swam, are of limited intelligence and possess fewer capabilities. The swarm of combined individuals is much more powerful than the individual members.
- Individuals in a swarm act and perform their respective tasks as per the rules and tasks allocated. The rules sometimes create complex behaviour as in bird flocks, where each individual stays within a certain distance to its neighbouring bird and flies slightly to the rear of it.

According to Bonabeau and Meyer (Bonabeau & Meyer, 2001) the following are the three main reasons of success of swarms:

1. Flexibility
2. Robustness
3. Self-organization

Ants are highly flexible living organisms in terms of adapting to changing environments, and ants soon find an optimal path between a food source and the nest if any obstacle arises.

Due to a large number of individuals in a swarm, the swarm has the characteristic of robustness. The swarm as a whole continues to exist, even if single individuals fail or become extinct.

Self-organization is a property of many complex systems that consist of a great amount of interacting subsystems. One characteristic of self-organization is a spontaneous forming of structures and patterns under specific conditions,wherefore it is suitable to start analyzing self-organizing processes by studying these so-called "pattern formations". In the honeybees' colony it is easy to find a multiplicity of such patterns. The organization in the hive arises from the partial systems as well, because the co-operating bees (the sub-systems) give structure to the system though they neither have knowledge of the global situation nor having a blueprint for the complete pattern in mind. Each individual works (blind for the holistic pattern) to the local rules, the global pattern results from the interaction of these single individuals. Every single bee has the ability to perceive only a little part of the entirety of influencing stimuli, but the whole of bees appears as a "collective intelligence".

3.4 Collective Intelligence of Natural Animals

The concept of "collective intelligence" is very hard to understand. In order to develop and innovate information systems, it is highly important to gain complete control over the environment. The term "collective intelligence" is related to "intelligence" as a phenomenon. In order to develop a theory of collective intelligence, it is highly desirable to understand the general terminology of intelligence.

A system is said to have "collective intelligence" when two or more "intelligent" sub-systems combine and interact with each other in an intelligent manner.

The word intelligence is closely related to intelligent behaviour. Intelligent behaviour is regarded as various actions and results in response to queries to ensure the system's reliability. Collective intelligence systems must have intelligent behaviour. The most obvious examples of collective intelligence are collective efforts of social organisms like ants, bees, fish, wasps, and termites when doing a particular task like collecting food, building nests, etc.

Collective intelligence has two dimensions: the *quality of intelligence* of individual swarm agents operating in an environment, which is also termed "bio-intelligence", and the *number* of swarm intelligence, which can range from billions to trillions, depending on the society. In addition to these, collective intelligence is also enforced by one more dimension, which involves a methodology of the systems integrated with and operating in an environment.

The concept of collective intelligence of an insect is not determined by its genetic behaviour, rather it emerges from continuous interactions of a large number of individuals that exhibit simple, stochastic behaviour and limited knowledge of the environment. Examples: the collective intelligence of ants – how the ants act as workers to locate the food dynamically in an environment, then come back to the nest, and other ants follow the trail from the nest to the food source in efficient manner. The collective intelligence of bees – the waggle dance, which provides the direction and distance from the nest of food, and its quality. The most important point is how chemical messages change the operating environment of insects in colonies, and how they repeat the behaviour to accomplish a particular task.

Therefore, collective behaviour is simply a sum of individual actions; it shows new characteristics that emerge from self-amplifying interactions among individuals and between individuals and an environment.

Another important aspect in collective intelligence is self-organization – how an individual swarm agent knows the work to perform without any coordinating authority.

3.4.1 Principles of Collective Intelligence

According to Don Tapscott and Anthony D. Williams (2008), collective intelligence is also termed "mass collaboration".

The following are the four principles of collective intelligence:

- Openness: The foremost principle of collective intelligence is sharing ideas. All the swarm agents share common information with their respective colonies, which provides them with an edge and more benefits from sharing information, and agents gain significant control and improvement via collaboration.

- Peering: Peering lays a strong foundation for "self-organization", a unique style of production that works more effectively than a hierarchy for performing all sorts of complex tasks like ant trailing, optimal route selection, etc.

- Sharing: With the sharing of ideas among swarm agents, such as the waggle dance of honeybees, the glowing capacity of fireflies, etc., the operations in the colonies become more effective and easy.

- Acting globally: With global integration, there exists no geographical boundaries, allowing swarm agents to operate freely in an environment and be more effective in collaborating with other agents for food search, trailing, and food collection from source to nest.

3.4.2 Characteristics of Collective Intelligence

The behaviour and work of colonies of social insects integrate many characteristics of collective intelligence and are listed here (Wolpert & Tumer, 1999):

1. Flexibility: The environment changes dynamically, so individuals should adapt to changing environmental conditions via an

amplification by positive feedback so that the entire system adapts to new circumstances.

2. Complexity: The entire colony behaviour adapts to an ever-changing environment, which could be quite complex in nature, and also involves solving the problem of coordination among different swarm agents to operate effectively in the environment.

3. Reliability: The system can function in an efficient manner even though some units fail and even without any central coordination utility to reassign duties in colonies.

The following are some of the real-time examples involving distributed intelligent computational systems, which require a centralized communication and can operate effectively using collective intelligence.

- Control systems for operating communication satellites (GPS satellites, weather forecasting satellites, or military satellites) for collecting and communicating scientific data.
- Control systems for effective routing of data in large communication networks (WAN or MAN).
- Parallel algorithms for solving complex optimization numerical problems such as scientific calculations, mainframe computing, or parallel systems for effective operations.
- Traffic control systems for large cities with an enormous number of vehicles operating at any point of day.
- Information grid control.
- Control of nuclear reactors or chemical plant for all sorts of logistics and control operations.
- Management of power electronics in large power grids based on wind mills, hydro power, etc.

3.4.3 Types of Collective Intelligence

The research on collective intelligence revolves around the investigation for the design of intelligent infrastructures that enable agents in colonies to think and act intelligently and intriguingly as compared to individual agents.

The following are the four different forms of collective intelligence which co-exist in natural animals and humans (Lévy, 1997):

1. Swarm
2. Original
3. Pyramidal
4. Holomidal

- Swarm collective intelligence

Swarm collective intelligence applies to a large number of individuals, from a few hundred to thousands to billions. Individuals don't have a big difference from one another via design or by context. Their individual margin of actions remains limited and in most of the cases remains predictable. Swarm collective intelligence can show immense adaptive and learning capacities.

Examples: various social insects (ants, bees, termites, wasps), schools of fish, flocks of birds, and even traffic jams.

- Original collective intelligence

Original collective intelligence consists of small groups composed of specialized individuals such as wolf swarm, dolphin swarm, cat swarm, monkey swarm, etc.

- Pyramidal collective intelligence

Pyramidal collective intelligence comprises a hierarchy-based intelligence structure, which provides the best coordination and effectively utilizes the power of the masses. It is mostly used in human organizations because of their civilization. In civilization, a human intelligence has defined criteria, like: urban concentrations, work specialization, accounting and memorizing, large geographical area, and centralized power.

Pyramidal collective intelligence has the following four dynamic principles (Bastiaens, Baumöl, & Krämer, 2010):

1. Labor Division: Everyone in civilization has a predefined role and can be interchangeable.
2. Authority: Determines the rules; assigns rights and prerogatives.
3. Limited Currency: Medium of exchange and a store of value.
4. Standards and Norms: Circulation and interoperability of knowledge within the community.

- Holomidal collective intelligence

It is one kind of intelligence that is at a developing stage and succeeds pyramidal collective intelligence as it has become powerless to address the tomorrow's high-stake complexity of intelligence. Holomidal collective intelligence is showing an enormous amount of growth in various forms of technical and scientific applications, and innovations in arts.

3.4.4 Difference between Social Intelligence and Collective Intelligence

Collective Intelligence, as mentioned above, highlights the two main points: First, there exist some species of natural animals, whose collective intelligence is regarded as much more sophisticated and intelligent than human beings'. Second, some animals, such as dolphins and monkeys, share a sense of group behaviour, i.e., social intelligence.

So, it is highly important to mention a clear distinction between two terms, "social intelligence" and "collective intelligence".

Social intelligence is regarded as an advanced form of intelligence, which makes an animal capable of performing complex behaviour patterns to analyze and respond to situations correctly with other members in the group. Social intelligence is regarded as a situation in which mammals behave in a way that is termed as "collective consciousness".

Collective intelligence is also regarded as an advanced intelligence, in which the group intelligence of individual swarm agents copes correctly with a social and physical as well as biological environment.

Ants have a higher degree of collective intelligence as compared to social intelligence. Social intelligence has a genetic capability to respond to all sorts of signals from conspecifics. The low degree of social intelligence of ants allows interlopers to confuse the ants completely. Example: Beetle begging behaviour causes ants to regurgitate droplets of food for them. The parasites can even perform the best mimicry of ants, not only in terms of behaviour but also with pheromone interference, to take food to their nests.

Low social intelligence can also be found in bee hives. Italian and Australian bees share almost the same signals and confuse each other.

Although, social insects that rank very low in social intelligence rank on the top in collective intelligence. Collective intelligence is mostly based on the mutual understanding and self-organization behaviour of agents.

3.5 Concept of Self-Organization in Social Insects

In recent years, the concept of self-organization (Bell & Cardé, 2013; Bonabeau, Theraulaz, Deneubourg, Aron, & Camazine, 1997) has been used to derive the emergence and maintenance of complex stable states from simple component systems. It was primarily introduced with a relevance to physics and chemistry to define how various microscopic processes give birth to macroscopic structures in out-of-equilibrium systems. The theory of "self-organization" can even be further enhanced to ethological systems like social insects and animals (ants, bees, bats, fish, etc.) to demonstrate how the collective behaviour, which is complex, can emerge from interactions among individuals that exhibit simple behaviours.

Self-organization (Uyehara, 2008) is regarded as a process in which complex patterns originate without any sort of directions internally or any sort of supervising source. The self-organization is observed in varied systems like phase-transitions, chemical reactions, body muscles, and schools of fish.

Self-organizing systems, with reference to biology, are composed of group of individuals working collectively to create complex patters via positive and negative feedback interactions. Various social insects' behaviour is derived from self-organization, as the social insects don't have a cognitive power to generate their observed complex structures and activities. Examples: ant foraging, honey bee thermoregulation, ant corpse-aggregation, etc.

Examples of self-organization in the real world: The example of sporting oarsmen in a rowing team working collectively by pulling up oars in perfect synchronization with each one another with proper adjustments and same frequency. The rhythmic contractions of muscle fibers in the heart is the best example of self-organization.

The concept of self-organization in physical systems differs from that in biological systems in two ways:

- Complexity of sub-units: In physical systems, the sub-units are inanimate objects. In biological systems, there is some greater inherent complexity as sub-units are living organisms like ants, bees, fish, etc.

- Rules governing interactions: In physical systems, patterns are created via interactions based on physical laws. Biological systems do have behaviours based on laws of physics, in addition to physiological and behavioural interactions among living components.

The following are the four constituents of self-organization in social insects:

- Positive feedback (amplification): Simple behavioural patterns that leads to a creation of structures. Examples: recruitment and reinforcement. In ants, a food source is determined randomly by ants searching in environment and then a trail is laid and is followed from a source to the nest; indicating the location of a food source by honey bees via the waggle dance.

- Negative feedback counterbalances positive feedback and stabilizes collective patterns like saturation, exhaustion, and competition. Examples: foragers, food source finish, food source crowding, etc.

- Self-organization is based on the amplification of fluctuations like random walks, errors, random task-switching, etc. Example: loss of foragers in an ant colony.

- All self-organization relies on multiple interactions.

3.6 Adaptability and Diversity in Swarm Intelligence

Swarm intelligence is the collective behaviour of decentralized, self-organized systems, natural or artificial (Bonabeau et al., 1999). The emergent collective intelligence of groups of simple agents is termed swarm intelligence. Individually, these agents are not intelligent but they have self-organization and division of labor in their nature. Collectively, these agents follow simple rules while interacting with others without any centralized control.

Adaptation and diversity in swarm intelligence-based algorithms can be seen as a proper balance between intensification and diversification, controlling and fine-tuning of parameters, and the identification of new solutions and the precise aggregation of uncertainty.

The success of a swarm intelligence-based algorithm depends on two contradictory process: exploration of search space and exploitation of best feasible solution. The first process is called diversification, and the second is known as intensification, both are driving forces for swarm intelligence. Most of the population-based techniques are diversification oriented, whereas single solution-based techniques are intensification oriented. The local search strategies lead to better intensification, and random search strategies lead to better diversification. The main criterion that one needs to deal with while initializing a population-based swarm intelligence strategy is diversification. A weakly diversified initial population leads to premature convergence. Initialization strategies with decreasing diversity capabilities are as follows (Talbi, 2009): parallel diversification, sequential diversification, the quasi-random approach, the pseudo-random approach, and the heuristic approach. Thus, it is essential to maintain a balance between intensification and diversification. These, swarm-based algorithms have some parameters that directly or indirectly control the diversity of a population. A fine-tuning of these control parameters is highly required along with adaption to the environment.

The crucial model of adaptation is a balance between intensification and diversification in swarm intelligence, and diversity is inherently associated with it. A problem-specific search process with strong exploitation leads to an intensive search and concentrated diversity that results in a rapid loss of diversity and subsequently premature convergence. While completely randomized process can enhance exploration and locate more diversified solutions. However a higher degree of diversity may mislead the search process and results in very slow convergence. That's why it is very crucial to make this search process adaptive. Most of the swarm-intelligence-based algorithms use a real number solution vector. The population size in these algorithms may be static or capricious. A varying size of population can maximize performance of algorithm. The adaptation has the capability to adjust algorithm-dependent parameters as a requirement of environment.

Likewise, a change in the radius of the search space, according to the iteration number, can lead to better search efficiency. This type of parameter tuning is known as a real parameter adaptation (Yang, 2015). Bansal et al. (2014) proposed a self-adaptive ABC with adaptation in the step size and limit parameter.

In the same way, the diversity in swarm-intelligence-based algorithms also takes various forms. Most of the swarm-based algorithms update a solution using specific equations. These equations have some random components and steps. These random components control the size of steps, and the performance of an algorithm depends on the step size. If step size is too large, then there are chances to skip true solutions, and if the step size is very tiny, then it may lead to a slow convergence. Thus, it is very crucial to maintain a proper amount of randomness. In order to maintain a proper diversification, Bansal et al. (2013) added a local search phase in ABC algorithm.

Adaptation and diversity in swarm intelligence are also related to the selection solution, i.e., which solution is to be forwarded for the next iterations and which one will be abandoned. Population-based algorithms perform this task using a fitness function. A highly fitted solution will be selected and the worst-fitted solution discarded during this selection process. In this case, a good adaptation policy is highly required can also maintain the diversity of solutions. A self-balanced particle swarm optimization developed by Bhambu et al. (2017) allows one to maintain a proper balance between the diversification and convergence abilities of the population. Xin-She Yang (2015) discussed adaptation and diversity, exploration and exploitation, and attraction and diffusion mechanisms in detail. It is not possible to study adaptation and diversity independently; they are interlinked with each other, and it is essential to retain an appropriate balancing between them.

3.7 Issues Concerning Swarm Intelligence

Swarm intelligence has become an important subject these days for addressing many issues in the research of engineering problems. It has wide applications in wireless communications, robotics, surveillance, biomedical applications, etc. As part of research it is very important to understand issues that arise while adopting the swarm intelligence principles and applying them to the subject considered.

The modern engineering applications adapt the swarm intelligence for improving the overall reliability of the system performance. Swarm intelligence is the technology in which many members are associated in order to obtain the prime objective of overall standardization of the system (Banzi, Pozo, & Duarte, 2011). Obviously, when many members are associated to each other, there are chances for security issues related to the information. The intruder's association with the swarm may cause serious issues in data communication between members and may corrupt the data (Dai, Liu, & Ying, 2011).

It is very important to think about the security challenges that may arise in the areas like communications, robotics, and cloud-computing-based Internet of Things in particular. Swarm intelligence is implemented based on the data accumulated from many swarm members associated in the network. So, because the information must be carried from one member to others, we should consider security. One should consider using a special type of encryption and decryption codes for encoding and decoding data that is communicated over the channels between swarm members; otherwise, an intruder might change or control the data.

In mobile sensor networks like MANET, a set of the devices or nodes is connected by a wireless network. The flying robots, such as drones, are associated with sensor networks used for some security applications and use swarm intelligence for the association and effective decisions. The communication in the preceding applications is through a wireless medium. Here, swarm intelligent networks go beyond the capabilities of mobile sensor networks because the latter are not capable of duplicating a developing behaviour of swarm intelligence networks.

The confidentiality of the system members and their data distributed in the swarm is becoming an important issue. The protection of the data is a first and foremost priority in the swam-intelligence-based applications. The next concern is to protect the integrity of the data. Data communicated among the members can be accessed only in authorized ways. Service availability is also an important issue for applications of swarm intelligence. Denial of service (DoS) is about losing the accessibility of service. It is very important that most of the security issues arise are same as we see in other systems. The resource constraints are more important in the case of swarm intelligence. The smaller the system that is considered, the tougher the security is for it. This is because of the resource constraints such as memory, communication bandwidth, power of computation, and energy. In the robotic swarm, if a certain attack on a particular robot happens, then the behaviour of an entire swarm may be affected. Suppose the robot is influenced and tampered with, then reintroduced into the swarm. An attacker may then be influencing the complete swarm. Controlling an object is a very sensitive issue in swarm intelligence. Because there is no central control in the swarm intelligence concept, members of the swarm are solely dependent on others' intelligence. The identity and authentication of the robot play a major role in the swarm intelligence. Different authentication schemes are in use such as group identity and individual identity.

Intruder identification in the swarm is very important. The intruder in the group can be identified easily. The abnormal behaviour of a member as compared to other members' behaviour shows the difference. Once the intruder is identified, it must be segregated from the system. Finally, a swarm member should have efficient learning ability.

3.8 Future Swarm Intelligence in Robotics – Swarm Robotics

Swarm robotics (Şahin, 2004; Brambilla et al., 2013; Gogna & Tayal, 2013) is regarded as a novel approach that is concerned with the coordination and cooperation of a large number of simple and powerful robots, which are autonomous, not even controlled centrally, capable of local communication and fully operational on the basis of the biological approach.

The main characteristics which make up a swarm robotic system are:

- Autonomous approach.
- Operational in specific environment with ability to modify themseleves as per changing environmental conditions.
- Sensing and communication capabilities in stored memory.
- No centralized controlling mechanism and global knowledge.
- Cooperation among other robotic systems to accomplish tasks in coordinated manner.

In order to elaborate further, regarding definition of swarm robotics, we can split an entity as follows:

- Machine: Entity having mechanical operational behaviour.
- Automation: Entity having capability of informational behaviour, i.e., information transfer as well as processing.
- Robot: Mechanical automation, i.e., an entity having both mechanical and information behaviour.

3.8.1 Advantages of Swarm Robotics

In order to understand the concept and advantages of swarm robotics in more deep manner, we compare advantages of swarm robotics with a single robot (Tan, & Zheng, 2013):

1. Accomplishing a Tedious Task: In order to accomplish a tedious task, single robot design and development can lead to a complicated structure and require high maintenance. A single robot can even face some problems with component failure and if, at any point of

time, a single component fails, the entire robot can stand still and task is halted. As compared to single robot systems, swarm robotics can perform the same task with a high agility in less time with error-free operation and with less maintenance.

2. Parallel Task Execution: A single robot can do one single task at a time like a batch processing system in computer. As compared to this, swarm robotics have high population size in which many robots can work at single point of time accomplishing multiple tasks in a coordinated manner.

3. Scalability: Swarms have local interaction in which individuals can join or quit a task at any point of time without interrupting the whole swarm. The swarm can easily adapt with environmental change and re-allocate units from the task to perform external operations.

4. Stability: As compared to single robot system, swarm robotics are much stable and reliable, and their work is even not affected when a single swarm quits work. The swarm robotic system will keep on working even with few swarms, but having few swarms will impact the overall performance and cause task degradation.

5. Economical: When building an intelligent self-coordinating swarm robotic structure, the cost will be much lower with regard to design, manufacturing, and maintenance as compared to single robot. Also, swarm robots can be produced in a mass manner, whereas a single robot development is a high-precision task and takes much more time.

6. Energy Efficient: As compared to single robot system, swarm robotics are smaller and highly simple in operation; therefore, the energy consumption cost is far less as compared to single robot taking into consideration the battery size. Swarm robots can operate for a large time, doing all sorts of complex tasks as compared to a single robot system, which requires battery recharging after a certain period.

Considering all the benefits of swarm robotics, as mentioned above compared to a single robot system, swarm robotics can be applied to high-end applications such as military applications, disaster relief operations, mining, geological surveys, transportation, and even environmental monitoring control.

3.8.2 Difference between Swarm Robotics and Multi-Robot Systems

The following Table 3.2 highlights the differences between swarm robotics and multi-robot systems:

TABLE 3.2

Swarm Robotics vs. Multi-Robot Systems

Points of Difference	Swarm Robotics	Multi-Robot Systems
Size	Large number, and variations can be done in real time.	Small as compared to Swarm Robotics.
Control	No centralized control; operate almost autonomously	Centralized control or remote-control operations
Flexibility	High	Low
Scalability	High	Low
Operational Environment	Almost unknown	Known and even operate in unknown environments
Homogeneity	Homogenous	Heterogeneous
Applications (Real-Time)	Military operations, mining, relief and environmental monitoring	Transportation

3.9 Conclusion

This chapter aims to provide a basic introduction to various concepts of swarm intelligence – metaheuristics and swarm. The chapter follows a discussion with regard to various advanced concepts of swarm intelligence, such as collective intelligence and self-organization as well as adaption and diversity. It also focusses on some issues concerning swarm intelligence and concludes with an advanced area of swarm intelligence application in robotics, i.e., swarm robotics.

References

Bai, H., & Zhao, B. (2006). A survey on application of swarm intelligence computation to electric power system. In: *Proceedings of the 6th World Congress on Intelligent Control and Automation*, Vol. 2, pp. 7587–7591.

Bansal, J. C., Sharma, H., Arya, K. V., Deep, K., & Pant, M. (2014). Self-adaptive artificial bee colony. *Optimization*, 63(10), 1513–1532.

Bansal, J. C., Sharma, H., Arya, K. V., & Nagar, A. (2013). Memetic search in artificial bee colony algorithm. *Soft Computing*, 17(10), 1911–1928.

Banzi, A. S., Pozo, A. T. R., & Duarte, E. P. (2011). An approach based on swarm intelligence for event dissemination in dynamic networks. In *30th IEEE Symposium on Reliable Distributed Systems (SRDS)*, pp. 121, 126, 4–7 October.

Bastiaens, T. J., Baumöl, U., & Krämer, B. (Eds.). (2010). *On Collective Intelligence* (Vol. 76). Berlin, Germany: Springer Science & Business Media.

Bell, W. J., & Cardé, R. T. (2013). Chemical ecology of insects. Berlin, Germany: Springer.

Beni, G. (1988). The concept of cellular robotic systems. In *Proceedings of the IEEE International Symposium on Intelligent Systems*, pp. 57–62. Piscataway, NJ: IEEE Press.

Bhambu, P., Kumar, S., & Sharma, K. (2017). Self-balanced particle swarm optimization. *International Journal of System Assurance Engineering and Management*, 1–10.

Bonabeau, E., Dorigo, M., & Theraulaz, G. (1999). *Swarm Intelligence: From Natural to Artificial Systems* (No. 1). Oxford: Oxford University Press.

Bonabeau, E., & Meyer, C. (2001). Swarm intelligence: A whole new way to think about business. *Harvard Business Review*, 79(5), 106–115.

Bonabeau, E., Theraulaz, G., Deneubourg, J. L., Aron, S., & Camazine, S. (1997). Self-organization in social insects. *Trends in Ecology & Evolution*, 12(5), 188–193.

Brambilla, M., Ferrante, E., Birattari, M., & Dorigo, M. (2013). Swarm robotics: A review from the swarm engineering perspective. *Swarm Intelligence*, 7(1), 1–41.

Civicioglu, P. (2012). Transforming geocentric cartesian coordinates to geodetic coordinates by using differential search algorithm. *Computers & Geosciences*, 46, 229–247.

Civicioglu, P. (2013). Artificial cooperative search algorithm for numerical optimization problems. *Information Sciences*, 229, 58–76.

Dai, Y., Liu, L., & Ying, L. (2011). An intelligent parameter selection method for particle swarm optimization algorithm. In *2011 Fourth International Joint Conference on Computational Sciences and Optimization (CSO)*, pp. 960, 964, 15–19 April.

Dorigo, M. (1992). Optimization, learning and natural algorithms. PhD Thesis, Politecnico di Milano, Italy.

Eberhart, R., & Kennedy, J. (1995, October). A new optimizer using particle swarm theory. In *Proceedings of the Sixth International Symposium on Micro Machine and Human Science, 1995* (pp. 39–43). IEEE.

Farmer, J. D., Packard, N. H., & Perelson, A. S. (1986). The immune system, adaptation, and machine learning. *Physica D: Nonlinear Phenomena*, 22(1–3), 187–204.

Glover, F. (1977). Heuristics for integer programming using surrogate constraints. *Decision Sciences*, 8(1), 156–166.

Gogna, A., & Tayal, A. (2013). Metaheuristics: Review and application. *Journal of Experimental & Theoretical Artificial Intelligence*, 25(4), 503–526.

He, S., Wu, Q. H., & Saunders, J. R. (2009). Group search optimizer: An optimization algorithm inspired by animal searching behavior. *IEEE Transactions on Evolutionary Computation*, 13(5), 973–990.

Holland, J. H. (1992). Adaptation in natural and artificial systems: An introductory analysis with applications to biology, control, and artificial intelligence. Cambridge: MIT Press.

Karaboga, D. (2005). *An Idea Based on Honey Bee Swarm for Numerical Optimization* (Vol. 200). Technical report-tr06, Erciyes university, engineering faculty, computer engineering department.

Kirkpatrick, S., Gelatt, C. D., & Vecchi, M. P. (1983). Optimization by simulated annealing. *Science*, 220(4598), 671–680.

Lenstra, J. K., Kan, A. R., & Shmoys, D. B. (1985). *The Traveling Salesman Problem: A Guided Tour of Combinatorial Optimization* (Vol. 3, pp. 1–463, E. L. Lawler, Ed.). New York: Wiley.

Krishnanand, K. N., & Ghose, D. (2005, June). Detection of multiple source locations using a glowworm metaphor with applications to collective robotics. In *Proceedings 2005 IEEE Swarm Intelligence Symposium, 2005* (pp. 84–91). IEEE.

Koza, J. R. (1994). *Genetic Programming II: Automatic Discovery of Reusable Subprograms*. Cambridge, MA: MIT Press.

Lévy, P. (1997). *Collective Intelligence*. New York: Plenum/Harper Collins.

Maniezzo, A. C. M. D. V. (1992). Distributed optimization by ant colonies. In *Toward a Practice of Autonomous Systems: Proceedings of the First European Conference on Artificial Life* (p. 134). Mit Press.

Mucherino, A., & Seref, O. (2007, November). Monkey search: A novel metaheuristic search for global optimization. In *AIP Conference Proceedings* (Vol. 953, No. 1, pp. 162–173). AIP.

Osman, I. H., & Laporte, G. (1996). Metaheuristics: A bibliography. *Annals of Operations Research*, 63, 513–562.

Şahin, E. (2004, July). Swarm robotics: From sources of inspiration to domains of application. In *International Workshop on Swarm Robotics* (pp. 10–20). Berlin: Springer.

Shah-Hosseini, H. (2011). Principal components analysis by the galaxy-based search algorithm: A novel metaheuristic for continuous optimisation. *International Journal of Computational Science and Engineering*, 6(1–2), 132–140.

Sörensen, K., Sevaux, M., & Glover, F. (2018). A history of metaheuristics. *Handbook of Heuristics*, 1–18. Springer

Sörensen, K., & Glover, F. W. (2013). Metaheuristics. In *Encyclopedia of Operations Research and Management Science* (pp. 960–970). Springer US.

Stonier, T. (2012). *Beyond Information: The Natural History of Intelligence*. Springer Science & Business Media.

Storn, R., & Price, K. (1997). Differential evolution–a simple and efficient heuristic for global optimization over continuous spaces. *Journal of Global Optimization*, 11(4), 341–359.

Talbi, E. G. (2009). *Metaheuristics: From Design to Implementation* (Vol. 74). John Wiley & Sons.

Tan, Y., & Zheng, Z. Y. (2013). Research advance in swarm robotics. *Defence Technology*, 9(1), 18–39.

Tapscott, D., & Williams, A. D. (2008). *Wikinomics: How Mass Collaboration Changes Everything*. New York: Penguin.

Uyehara, K. (2008). Self-organization in social insects.

Wolpert, D. H., & Tumer, K. (1999). An introduction to collective intelligence. *arXiv preprint cs/9908014*.

Yang, X. S. (2010). Firefly algorithm, Levy flights and global optimization. In *Research and Development in Intelligent Systems XXVI* (pp. 209–218). Springer, London.

Yang, X. S. (2010). A new metaheuristic bat-inspired algorithm. In *Nature Inspired Cooperative Strategies for Optimization (NICSO 2010)* (pp. 65–74). Springer, Berlin, Heidelberg.

Yang, X. S. (2015). Nature-inspired algorithms: Success and challenges. In *Engineering and Applied Sciences Optimization* (pp. 129–143). Cham: Springer.

Yang, X. S., & Deb, S. (2009, December). Cuckoo search via Lévy flights. In *World Congress on Nature & Biologically Inspired Computing, 2009* (pp. 210–214). IEEE.

4

Ant Colony Optimization

Bandana Mahapatra and Srikanta Patnaik

CONTENTS

4.1 Introduction

Ant colony optimization (ACO) is a part of swarm intelligence that mimics the cooperative behaviour of ants to solve hard combinatorial optimization problems. Its concept was pioneered by Marco Dorigo (1992) and his colleagues who were astonished when observing the inter-communication and self-organized cooperative method adopted by these creatures while searching for food. They conceived the idea of implementing these strategies to provide solutions to complex optimization problems in different fields, gaining a lot of popularity (Dorigo, 1992; Papadimitriou & Steiglitz, 1982).

ACO is a set of software agents called artificial ants, who look for good solutions to specific optimization problems. The ACO is implemented by mapping the problem into a weighted graph, where the ants move on the edges in search of the best path.

ACO has come a long way since its conceptualization in year 1992 by Marco Dorigo in his PhD thesis (Dorigo, 1992). Before proceeding to the concept of ACO, it is necessary for us to honor the contribution of researchers in this field.

The study of ACO, or in fact real ants, came to existence in 1959, when Pierre Paul Grasse invented the theory of "stigmergy," explaining the nest building behaviour of termites. This was followed by Deneubourg and his colleagues in 1983 (Deneubourg et al., 1991), when he conducted a study on the "collective behaviour of ants." In 1988, Mayson and Manderick contributed an article on the self-organization behaviour observed among ants. Finally, in 1989, the research work of Goss, Aron, Deneubour, and Pasteelson (the collective behaviour of Argentine ants) (Deneubourg, Aron, Goss, & Pasteels, 1990) produced the base idea of the ACO algorithm. In the same year, Ebling and his colleagues proposed a model for locating food, and in 1992, the Ant System was proposed by Marco Dorigo in his doctoral thesis that was published in 1997 (Dorigo, 1992). A few researchers extended these algorithms to applications in various areas of research. Appleby and Steward of British Telecommunications published the first application in telecommunication networks, which was later improved by Schoonderwoerd and his colleagues in 1997. In 2002, it was applied for scheduling problems in a Bayesian network.

The ACO algorithms designed are based upon ants' capability to search for short paths, connecting the nest and food location, which may vary from ant to ant, depending upon its species. In recent trends, researchers have conducted a study on the results obtained from applying ACO that suggests that majority of the artificial ants employed do not provide the best possible solution, whereas elitist ants provide the best solution, undergoing a repeated swapping technique. They further state that the hybrid and swapping approach outperformed the Ant Colony System (ACS) provided the same parameters are considered.

On the basis of this theory, researchers have proposed a specialized ant colony optimization method that mimics specific species of ants, such as *Lasius niger* or *Irodomyrex humilis*, famously known as Argentine ants.

Apart from being a part of the solution to common routing and scheduling problems, ACO has also been implemented in unlikely areas. A few such contributions are applications to protein folding in 2005 and peptide sequence design in 2016, (a few more applications are discussed in Section 4.11). Rigorous research work is being continuously carried on, which will contribute to new revolutions in the upcoming era of science and technology.

In upcoming sections, the chapter will elaborate on the detailed working of ACO, its algorithms and the mathematical background, as well as its application areas.

4.2 Concept of Artificial Ants

The cooperative behaviour of natural creatures has often proved a great solution to solve a variety of complex problems in real-life scenarios. The collaborative behaviour of the living beings has attracted the researchers to incorporate their type of intelligence into artificial agents that work together toward solving various complex problems in diverse fields. This section will discuss the details of artificial ants (Dorigo, 1992).

4.2.1 How Real Ants Work

Tiny natural creatures like ants can work jointly toward dealing with complex tasks. This communal behaviour is based on special chemicals secreted by the ants when they are in motion, called pheromones.

These pheromones attract other ants to use the same path, which is the driving force that justifies why ants always communicate in colonies, how ant hills are created, and how they are able to achieve a complex structure through an intercommunication strategy.

As we discussed, the pheromone plays a significant role in exchange of information among the ants accomplishing various collective assignments; it becomes important for us to understand the working concept of pheromones and how they help the ants in finding the shortest path to the food source.

The capability of the ants to come up with strategies for solving complex problems has become a popular topic and is widely applied as an optimization technique to solve various research problems. Their unique skill of moving from "chaos to order" has become the major factor in drawing researchers' attention (Fogel et al., 1966).

4.2.2 How Ants Optimize Food Search

The food search process can be organized roughly into three stages. The ants start for their goal by walking in a random direction in search of food. These ants wandering in a chaotic fashion finally return to their nest when exhausted to eat and rest. However, while rambling, if one of them comes across a food source, it takes a tiny piece along with it, back to its nest, leaving a trail of pheromone. This pheromone acts as an indicator to the other ants regarding presence of food; as a result, the ants follow the pheromone depositing even more pheromone trail. However, considering the multiple path options available for the ants to reach the food source, the initial travel of the ants is somewhat chaotic in nature, with many paths connecting the nest and the food. Ants generally choose the pheromone path that is stronger. The pheromones evaporate with time, leaving the shortest path to reach the food with the strongest pheromone deposit; the ants slowly converge toward this route, making it the optimal one to be followed, which later emerges as an ant colony (Fogel et al., 1966).

4.2.3 Concept of Artificial Ants

The artificial ants are nothing but the simulating agents that mimic a few of the behavioural traits of real ants to solve a complex real-world problem. According to computer science, they represent multi-agent techniques derived from the real ants' behaviour. This concept is built on the idea, "intelligence can be a collective effort of many minuscule objects." This "ambient network" of intelligent individuals is currently the vision of future technology, which is expected to outperform the current centralized system based on the human brain (López-Ibáñez, Stutzle & Dorigo, 2006).

4.2.3.1 The Artificial Pheromone System

The pheromone-based interaction among the real ants is one of the attractive features possessed by these creatures. This intercommunication behavioural aspect has been implemented in varied forms (chemical, physical, light, audio, etc.) for intensifying the communication mode among the individual agents participating. The pheromone strength after pheromone trail evaporation is a driving factor that forces all ants to converge into most optimal paths out of other suboptimal ones available in search space. As we know, the exploring feature of the ants is mainly a contribution of the pheromone intensity, which varies from one path to another. The pheromone trail evaporation occurs at an exponential speed.

This concept can be implemented into a network system with the search path consisting of the nodes connected through arcs. The pheromone trail evaporation can be calculated as:

$$\tau_{ij} \leftarrow (1-\rho)\tau_{ij}, \ \forall (i,j) \in A \tag{4.1}$$

after an ant moves from one node to next node.

Here $p \in (0, 1]$ is the parameter after the pheromone evaporation occurs; as shown in equation 4.1, the amount of pheromone added to the arc path is Δe^k.

The act of pheromone evaporation helps the process of optimum path exploration to improve by forgetting the old paths, mainly those that are non-optimal or error based, which helps these artificial creatures to seek the maximum optimum value achievable by the pheromone trail (Dorigo & Stutzle, 2006).

4.3 Foraging Behaviour of Ants and Estimating Effective Paths

The visual perception of many ant species is poorly developed with a few species being completely blind. The pheromone secreted from their bodies acts as an insight of communication among the individuals. This makes the whole of the ant environment chemical based, which is different from the interaction with the environment by higher species such as human or aquatic animals, whose most dominant senses for communication are visual or acoustic. The pheromone trail, as discussed, is a prominent factor in maintaining the social life of the ant species. The trail of pheromone is a distinct variety of pheromone secreted by *Lasius niger* or *Iridomyrmex humilis* (Argentine ant) species (Goss, Aron, Deneubourg, & Pasteels, 1989), which is specifically used for marking paths on the ground. Foragers sense these pheromone trails to reach the food discovered by other ants. This trail-following behavior by the ants is the foundation on which ACO is built (Deneubourg et al., 1990).

4.3.1 Double-Bridge Experiments

As we know, many ant species such as *I. humilis* (Goss et al., 1989), *Linepithema humile*, and *Lasius niger* (Bonabeau, Sobkowski, Theraulaz, & Deneubourg, 1997) show foraging behaviour based upon an indirect communication medium that is built upon a pheromone trail connecting food source to the ant nest.

Several researchers have explored the ants' behaviour regarding the pheromone trail by conducting controlled experiments. Such an experiment was conducted by Deneubourg and colleagues (Deneubourg et al., 1990; Goss et al., 1989), where they used a double bridge, connecting the food source with the nest of the Argentine ant species, *I. humilis*. The experiment was conducted for varying ratios between the two branches of the bridge (Deneubourg et al., 1990).

$$\text{Ratio } (r) = L_l/L_s,$$

Where, L_l = Length of longer branch
L_s = Length of shorter branch

They started with the experiment conducted with both branches of equal length (i.e., $r = 1$) as shown in Figure 4.1(a). The ants were left to move around freely, choosing either of the branches to reach the food source. The percentage of ants choosing each branch was recorded over time. The result showed ants choosing one of the branches initially and all ants converging to the same one, as shown in 4.1(b).

The repetition of the same experiment was conducted with branches of varying length ($r = 2$) (Gross et al., 1989); the long branch was double the size of the smaller one. The percentage of ants choosing individual branches was recorded. The experimental outcome shows that over time all ants converge to the shorter branch, as shown in Figure 4.1(b). The result for the first experiment, shown in Figure 4.1(a), was justified because the ants started their search and chose randomly either of the branches because both appeared similar to them with no pheromone deposits. The probability of either branch being selected remains the same because they do not have any preference. Due to random fluctuation, a few more ants may select one branch over the other, depositing more pheromone and stimulating other ants to choose the same branch. Finally, the ants converge to a single path.

This self-organizing behaviour of ants serves as an example of ants being auto-catalystic or having positive feedback behaviour, which is the result of the interactions that take place at the microscopic level (Camazine et al., 2001). Here, the ants' behaviour, which shows convergence to single path, might also be termed stigmergic communication achieved by exploiting indirect communication mediated by environmental modifications where they move.

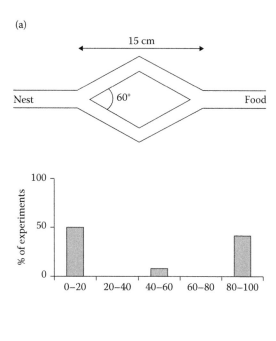

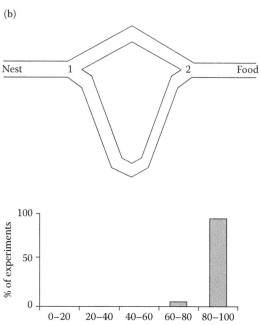

FIGURE 4.1
(a) Experiment with branch of same length and (b) experiment with branches having different lengths.

The resultant graph for the second experiment conducted, shown in Figure 4.1(b), can be explained as follows. Since both branches appear identical to the ants, they choose randomly. Hence, it can be assumed that half the ants choose the short branch, whereas half choose longer. The ants choosing the short branch reach food faster and start to return to their nest, depositing more pheromone than the ants using the other branch; thus, they attract other ants to choose the same path. The pheromone eventually accumulates faster on the shorter branch, which subsequently would be used by all ants due to auto-catalyst process described previously.

Deneubourg and colleagues conducted an additional experiment, where at the first instance a long branch of 30 Min was provided, as shown in Figure 4.2, and later on, after convergence, a shorter branch was introduced. The result of this experiment proved the sporadic selection of short branch, where the long branch remained trapped under the colony. This result can be accounted for by the high concentration of pheromone on the long branch as well as the slow rate of pheromone evaporation, which makes the majority of the ants choose the long branch on account of auto-catalytic behaviour, even after the shorter branch appears (Goss et al., 1989; Deneubourg et al., 1990).

Later a stochastic model describing the dynamics of the ant colony based upon the observed behaviour in double-bridge experiment was proposed (Goss et al., 1989; Deneubourg et al., 1990, 1991).

The metrics of the proposed model are given as:

Ψ – Ants per second crossing the bridge in each direction at a constant speed of v cm/s, depositing 1 unit of pheromone on the branch.

L_S & L_L (in cm) Length of the long branch & short branch

$t_S = L_S/V$ The time taken by an ant to traverse the short branch in seconds.

$r.t_S$ = Time taken to traverse the long branch in seconds

where,

$r = l_1/l_S$.

The probability $P_{ia}(t)$ for the ant to arrive at a decision point $i \in \{1,2\}$ and select branch $a \in \{S,L\}$

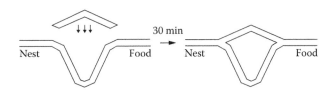

FIGURE 4.2
Experiment on ants with new path introduced after optimal path is formed.

(S = short branch, L= long branch) at an instant t can be represented as a function of total amount of pheromone on the branch

φ_{ia} (t) $\propto$ no of ants using the branch until time (t).

$$P_{is}(t) = (t_s + \varphi_{is}(t))^{\propto}/t_s + \varphi_{is}(t))^{\propto} + t_s + \varphi_{il}(t))^{\propto} \qquad (4.2)$$

Where, $P_{is}(t)$ = Probability of choosing short branch

$\propto$ = 2 (Neneubourg et al., 1990)

Similarly,

P_{il} (t) is computed using the relation P_{is} (t) + P_{il} (t) = 1

As we discussed in the previous section, this model considered the amount of pheromone on a branch that is directly proportional to the number of ants that used the branch at past and has ignored the pheromone evaporation.

Hence the differential equations describing stochastic systems evolution are given as:

$$d\varphi_{is}/dt = \Psi P_{JS} (t-t_s) + \Psi P_{is}(t) \; (I = 1, j = 2; I = 2, j = 1) \qquad (4.3)$$

$$d\varphi_{il}/dt = \Psi P_{Jl} (t-r \cdot t_s) + \Psi P_{il}(t) \; (I = 1, j = 2; I = 2, j = 1) \qquad (4.4)$$

where:

$d\varphi_{is}/dt$ = Instantaneous variation at time t of pheromone on branches at decision point i

Ψ = Ant flow (assumed constant)

P_{js} = Probability of choosing the short branch at decision point j.

$t - t_s$ = time

P_{is} = Probability of choosing the short branch at decision point i at time t.

t_s = Time delay (constant), i.e., the time needed for the ants to traverse the short branch.

i.e., the time needed for the ants to traverse the short branch.

4.4 ACO Metaheuristics

The challenge of solving combinatorial optimization problems can be considered fascinating for a researcher since majority of problems arising are practically NP hard, i.e., their optimal solution cannot be achieved within a given polynomial time frame. Hence, to solve such problems the target set is the near-optimal solution, achieved within a relatively short time span, using approximation methods. These kind of algorithms are termed heuristics that use problem-specific knowledge to either build or improve the solution. The current researchers have diverted their attention to a new category of algorithms, metaheuristics.

Metaheuristics may be stated to be a set of algorithmic concepts used to define heuristics methods that can be applied to solve a wide set of contrasting problems. They can be considered as a general-purpose heuristic method designed to guide the underlying problem-specific process. Heuristics often direct one toward a promising region of search containing high-quality solutions. Hence, metaheuristics are a general algorithmic framework, which can be applied to a variety of optimization problems of any specific kind, demanding few modifications for adaptability (Dorigo & Stutzle, 2006).

A few such examples are: tabu search (Glover, 1989, 1990; Glover & Laguna, 1997) iterated local search (Lourenço, Martin, & Stützle, 2002), evolutionary computation (Fogel, Owens, & Walsh, 1966; Rechenberg, 1973; Holland, 1975 Schwefel, 1981; Goldberg, 1989), and ant colony optimization (Dorigo & Di Caro, 1999; Dorigo, Di Caro, & Gambardella, 1999; Dorigo, Maniezzo, & Colorni, 1996; Dorigo & Stutzle, 2002; Glover & Kochenberger, 2002).

4.4.1 ACO Metaheuristics

Ant colony optimization (ACO) is a metaheuristic whereby a colony of artificial ants collectively find a good solution to difficult and discrete optimization problems. The key design component adopted by the metaheuristics to solve any of the discrete optimization problems is cooperation. Here, the target is distribution of computational resources within a set of designed artificial agents that use stigmergy as the medium of communication that is mediated by the environment. A good solution for these kinds of problems can be achieved as an outcome of the cooperative interaction of these agents.

The ACO algorithms are capable of solving both static and dynamic of combinatorial optimization problems.

The static problems are specified by their characteristics while defining the problem, which remains invariant while being solved. For example, the travelling salesperson problem (Johnson & McGeoch, 1997; Lawler et al., 1985; Reinelt, 1994), which is explained in Section 4.5 (Johnson, & McGeoch, 1997).

In contrast, dynamic combinatorial problems are defined as functions of the values set by the underlying system. Here, during the run time, the problem instance changes, hence the optimization algorithm must be capable enough to adapt itself to the change in environment. Examples of dynamic combinatorial problems are network routing problems with volatile network topology and data traffic that changes with respect to time.

4.4.1.1 Problem Representation

Although the ACO metaheuristic is applicable to any kind of combinatorial optimization problem, the major challenge is the mapping of the problem under consideration to a representation that can be used by the artificial ants for solution building.

This section provides a formal characterization of the representation the artificial ants used along with the policy they implement (Dorigo & Stutzle, 2006).

Considering the minimization problem (S, f, Ω)

Where,

S = Set of candidate solutions

f = Objective function that assigns cost value f (s, t) to each of the candidate solution, $s \in S$

$\Omega(t)$ = Set of constraints, where t is time

Here, the aim is finding a minimal cost feasible solution s^*.

The combinatorial optimization problem (S, f, Ω), is mapped on a problem, denoted by the list of symbols defined below.

- $C = \{c_1, c_2, c_3 \ldots \ldots, c_{nc}\}$, where C is a finite set of components, ranging from $1 \ldots \ldots \ldots .N_c$
- $X = < c_i, c_{j\prime\prime} \ldots \ldots \ldots, c_n \ldots \ldots >$, the sequence of finite length over elements defining the state of the problem
- $S \subseteq \chi$, where S is the set of candidate solutions
- $\tilde{x}$ is the set of feasible states, where $\tilde{x} \subseteq X$.
- S^* is a non-empty set of optimal solutions, where $S^* \subseteq \tilde{x}$ **and** $S^* \subseteq S$
- G (s, t) cost attached to each candidate solution $s \subseteq S$.

4.4.1.2 Ants' Behaviour

As explained in the previous sections, the artificial ants of the ACO algorithms are stochastically constructive procedures that are built solutions while moving on the construction graph G_C (C, L) where,

C is the component, and L is the connection within components.

$\Omega(t)$ is the problem constraints built into the ants' constructive heuristic.

Each ant (k) moving in the colony carries the following properties:

- It can use the constructed graph $G_c = (C, L)$ in searching for optimal solutions, $s^* \in S$.
- It possesses memory M^k that acts as storage for the information regarding the path covered so far. The presence of memory can be useful for:
    i.   Building feasible solutions (i.e., implementing constraints, Ω)
    ii. Computation of heuristic values η
    iii. Evaluating the solution obtained
    iv. Backtracking the path visited

- It begins with a start state x^k, which may be expressed as either empty or a unit sequence and ends with either one or more terminating conditions e^k.
- At state $x_r = <x_{r-1}, \ldots \ldots .i>$, if no termination condition is satisfied, the ant moves to state $<x_r, j> \in x$, i.e., node j, at neighbour $N^k(x_r)$. The ant is stopped from movement as soon as one of the terminating conditions e^k is satisfied. The ants are prohibited from entering the infeasible state while building the candidate solution due to a certain defined heuristic value η or given the ants' memory.
- The ants use the probabilistic decision rule to choose their next movements, where probabilistic decision rule is defined as the
    1. Given heuristic values and locally available pheromone trails associated with graph G_c.
    2. The current status of the ants stored in its memory.
    3. The problem constraint.
- Component C_j added to the current state updates the pheromone trails (τ) associated to it or its correspondent connection.
- Backtracking the same solution path is possible (Dorigo & Stutzle, 2006).

4.4.2 ACO Metaheuristics Problem Solving

The metaheuristic designed for ant colony optimization consists of three procedures:

1. Constructing the ant solution.
2. Updating the pheromone.
3. Implementing the centralized action also called the daemon action.

Construction of Ant Solution: This procedure consists of a colony of ants that contemporaneously and asynchronously or series-wise visits the adjacent state of the specified problem by visiting the neighbour nodes of the constructed graph G_C. The process of moving from a node to another is done by choosing the node for the next hop after applying the stochastic local decision policy, making use of the available heuristic information and the pheromone trails. Using this process, ants progressively build solutions to the optimization problems. During the building process, the ants evaluate (partial) solutions, which can be significant to the pheromone update procedure deciding the amount of pheromone to deposit.

   Pheromone update: Is the process of modification of the pheromone trail. With the deposition of pheromones by ants on components, or the connections used, the trail value increases; in contrast, it decreases on account of pheromone evaporation. Practically, the increase in the rate of pheromone deposited increases the probability of the path being used either by many ants or by at least a single ant that produces a good solution to be used by many ants in the future.

The phenomenon of pheromone evaporation is important for avoiding the ants' convergence to the suboptimal path with respect to the concept of forgetting, thereby encouraging them to explore new areas of search space.

Daemon actions: This procedure is used for implementation of the centralized action that is not possible on the part of the single ants. An example of this procedure is global information collection, which acts as the basis of the decision making regarding whether or not to add extra pheromone influence to the search procedure from a nonlocal one. As a practical example of this procedure, the daemon studies the path found by each participating ant in the colony, and the ants contributing toward building the best solutions during algorithm iteration are allowed to deposit additional pheromone on the components/connection used in the path (Dorigo & Stutzle, 2006).

4.4.3 The ACO Metaheuristic Algorithm

Start
Step 1. Start build Ant solution ← updating partial solution
Step 2. Components and connections update (pheromone trail)
Step 3. Daemons Actions
        i.e., choose best solution out of all the solutions provided by Ants
     Go to step 2
End

4.5 ACO Applied Toward Travelling Salesperson Problem

The Travelling Salesperson Problem is a question faced by a traveller while choosing among a set of cities to be visited, planning the shortest possible trip, so as to visit each city just once before finally reaching home (Jaillet, 1988; Jacobs & Brusco, 1995).

This problem can be represented as a complete weighted graph $G = (C, P)$

Where, C represents cities to be visited and
$c = |C|$ number of cities visited by nodes.
P = The arch connecting the nodes C.
Each arch $(i,j) \in p$ is associated with a weight d_{ij} representing the distance between the city i and j.

The TSP can be considered as the challenge of searching for a minimum length Hamiltonian circuit in the graph where the Hamiltonian circuit is a short tour that goes through each node representing Graph G exactly once. The graph may be considered as symmetric, i.e., $d_{ij} = d_{ji}$, where the distance travelled remains same between cities irrespective of direction of the arcs travelled across.

The problem of TSP can be represented as a cyclic permutation of cities showing significance on relative order rather than positioning of cities (Jünger, Reinelt, & Thienel, 1994).

The TSP problem can be characterized by the following.

4.5.1 Construction Graph

The construction graph consists of a set of components C representing the set of nodes ($C = N$) and a set of arcs $p(p = A)$, where each arc carries a weight corresponding to the distance d_{ij} between the nodes it connects. Here, the different states of the problem represent the set of possible partial tours (Jacobs & Brusco, 1995).

4.5.2 The ACO Algorithms for the TSP

As mentioned, all proposed design ACO algorithms for TSP have a graph with nodes, arcs, and pheromone trails. The pheromone trails associated with the arcs result in a desirability factor, referred to as z_{ij}, whereby ants visit node j after node I, and $\eta_{ij} = 1/d_{ij}$ is the heuristic information chosen, i.e., the desirability of going from node i to j is inversely proportional to the distance between the cities under consideration. The pheromone trail is represented as metrics containing τ_{ij}'s as the element. Figure 4.3 shows the ant's probability of choosing its next hop (Knox, 1994).

The ACO Algorithm for solving the TSP is given as
$C = \{c_1, c_2, c_3 \ldots \ldots \ldots \ldots, c_n\}$
$A = \{a_1, a_2, a_3, \ldots \ldots \ldots \ldots, a_n\}$
Start
Step 1. Choosing City C_o to start the travel.
                for (i=1; i<= n; i++)
Step 2.   City ← ant a_i (position of an ant in graph)
Step 3.   C= 0
Step 4.   For (C_i = a+1; $C_i \neq C_a$; i++)
     Choose next city using probabilistically using pheromone deposit & heuristic values

Step 5.   tour construct ← C_j
Step 6.   if (i=n)
Step 7.   i=0

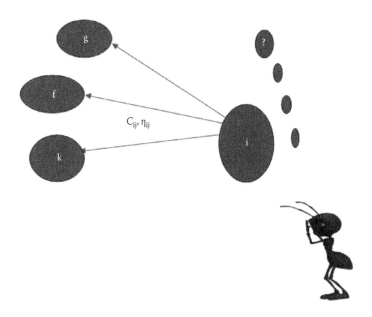

C_{ij}, η_{ij}

FIGURE 4.3
Ant determining the next hop among available nodes.

4.6 ACO Framework

This section discusses the standard framework of an ant colony optimization technique. The framework for ant colony optimization starts with the parameter initialization, where the various parameters are initialized. Following this step, the next move to the node in problem graph is decided, after which the tour is undertaken until all ants gets exhausted (Romeo & Sangiovanni-Vincentelli, 1991). On every tour, the tour quality is recorded and updated in the pheromone table. The process continues until the target is achieved, i.e., the number of iterations specified is completed. This generalized framework can be modified for solving various problems. Figure 4.4 shows the generalized ACO framework for solving problems.

The different improvised designs of the ACO framework proposed are:

1. Parallel ACO framework
2. Hypercube framework
3. Cognitive framework

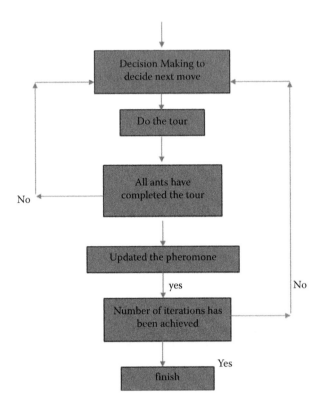

FIGURE 4.4
The generalized ACO framework.

4.6.1 Parallel ACO Framework

Here the methodology adopted for pheromone deposit calculation during the construction phase varies from the regular ACO framework. This theory was introduced with the goal to enhance the exploration and the exploitation capability of the artificial ants. In this theory, the exploration and exploitation techniques are inherited in the decision-making process. The advantage of this framework is that low probabilities are not neglected, encouraging more exploration than is achieved in the general ACO framework (Maniezzo & Milandri, 2002).

4.6.2 Hypercube Framework for ACO

In 2001, Blum, Roli, and Derigo introduced the concept of the hypercube framework, which rescales the pheromone values such that they lie within the interval [0,1]. This concept was based on the mathematical programming formation for combinatorial optimization problems where the solution possibly can be represented by binary vectors (Blum, Roli, & Dorigo, 2001).

The decision variable, assuming values {0,1}, corresponds to solution component the ants use to construct the solution. The solution to a defined problem corresponds to a corner of the n-dimensional hypercube, where n represents the possible decision variables. The lower bounds for any problem can be generated by relaxing the problem being considered that would facilitate each variable taking a value in the [0,1] interval. Here, S_{rx} can be defined as:

$$V \in S_{Rx} \Leftrightarrow V = \sum Y_i \cdot x_i \qquad (4.5)$$

$$\vec{V} \in S_{Rx} \Leftrightarrow \vec{V} = \sum_{X_i \in B^n} Y_i \vec{V} \qquad (4.6)$$

Here, S_{Rx}, is the set of all feasible solution.

$\vec{V} \in R^n$ is the set of all vectors $\vec{v}$ that are defined as convex combinations of binary vectors and $\vec{x} \in B^N$ where B^N is the set of binary vectors.

After normalization of the pheromone value to fit into the [0,1] interval, its association with ACO becomes explicit.

Here, the pheromone vector $\vec{\tau} = (\tau_1, \tau_2, \tau_3 \ldots \ldots, \tau_4)$ is found consistent to a point in $\vec{S}$ and to a solution if $\vec{\tau}$ is a binary vector.

4.6.3 Cognitive ACO Framework

The current ACO framework uses the sequential method of problem solving. Sequential processing runs in cycles. An extension of the sequential implementation has been proposed by researchers where the processing units have been split and executed parallelly. The concept behind this approach is to have reusable framework for several sequential algorithms in a parallel environment (Blum & Dorigo, 2004).

4.7 The Ant Algorithm

Ant algorithm may be defined as a multi-agent system inspired by the behaviour of real ants, which are capable of exploiting stigmergy (Dorigo & Stutzle, 2006). Apart from the foraging behaviour of ants, (discussed in previous section), there are other aspects of social behaviour that have inspired the computer scientists, such as brood sorting and division of labor. A few of the other aspects contributing to models of the ant algorithm are:

- Models inspired by foraging and path marking.
- Models inspired by brood sorting.
- Models inspired by division of labor (Handl & Meyer, 2002).

4.7.1 Foraging and Path Marking

Foraging, which can be described as path marking behaviour, has also been derived from ants to solve various real-time problems. Here, the artificial ants walk along the arcs of the graphs as well as lying pheromone over them (Wagner, Lindenbaum, & Bruckstein, 1998).

The models described in relation to this have the artificial ants walk on the arc and lay a pheromone trail on them. It uses the previously deposited pheromone trails to direct their exploration. The concept appears similar to ACO, but they are designed to achieve graph coverage in the absence of its graphical topology laid down by the ants' pheromones. This ant algorithm may be applied to graphs that are dynamic in nature.

4.7.2 Models Inspired by Brood Sorting

The brood sorting activity can be seen in a variety of ant species such as *Lasius niger* and *Pheidole pallidula*, where the ants cluster their small eggs, as well as micro-larvae, at the center of nest brood area, where the large larvae are present at the periphery of the brood cluster. A model was proposed by Deneubourg et al. (1991); the artificial ants mimic picking up and dropping items in their surroundings, i.e., the number of items picked and dropped is based on it. For example, an ant carrying a small egg would drop it in the region having a high population of small eggs, or vice versa if an ant found large larvae surrounding small eggs, there would be a high probability it will pick up the larvae.

Lumer and Faita (1994) and Kuntz Snyer and Lyzell (1999) found the model suitable for solving clustering problems (Lumer & Faieta, 1994). The model was also found suitable for robotics systems, being capable of building a cluster of objects with no centralized control or using robots to sort objects, as proposed by Holland and Melhuish (1999).

4.7.3 Models Inspired by Division of Labor

Among ants in an ant colony, there are certain ants that carry out specialized work, i.e., they are capable of adapting their behaviour in accordance to the circumstances. The combination of specialization and flexibility tends to be very appealing for multi-agent optimization and control, specifically in tasks related to resource allocation, which require continuous adaptation to environmental changes. Based on this behavioural concept Robinson (1992) proposed a threshold model where workers with a low response threshold respond to a low stimuli level in comparison to workers with a high response threshold. In the proposed model, stimuli is replaced with a stigmergic variable.

This model inspired by division of labor is mainly useful for solving problems related to dynamic task scheduling (Bonabeau, Dorigo, & Theraulaz, 1999, 2000; Bonabeau et al., 1997).

4.7.4 Models Inspired by Cooperative Transport

The ant colonies' behaviour has also inspired roboticists who were then interested in designing a set of distributed control algorithms for a group of robots (Martinoli & Mondada, 1998) engaging in cooperative box phishing. Kube and Zhang (1994) proposed a task used as a benchmark for ant algorithms. The concept here is that, when a single ant fails to retrieve a large item, nestmates or other ants are recruited to help through either direct contact or chemical marking, thereby implementing a form of cooperative transport. The ants keep moving the object they aim to carry, trying different positions and alignment until they achieve success in carrying it to their nest. Kube & Zhang have experimented with and achieved success in an ant algorithm where the behaviour of real ants has been reproduced in a group of robots whose aim is to push a box toward a goal (Kube & Zhang, 1994).

4.8 Comparison of Ant Colony Optimization Algorithms

As we know, ACO is a probabilistic technique formulated to provide a solution to various computational problems that aims to find the most favorable path through graphs. It has been diversified to solve an extensive class of numerical problems, giving rise to various other algorithms drawn from different behavioural aspects of ants.

A few of the common extensions of ACO are –

A. Ant system
B. Ant colony system
C. Max-min ant system
D. Rank-based ant system
E. Continuous interacting ACO algorithms
F. Continuous orthogonal ACO algorithms

4.8.1 Ant System

The ant system is considered a pioneer of all the ACO algorithms, since all proposed algorithms of this genre are based on the ant system. Here, the ants make a complete excursion of the considered graph, where the next node to be visited is chosen on the basis of certain probabilistic rules, and each iteration level updates the amount of pheromone deposited in the problem space. The amount of pheromone is said to be inversely proportional to the amount of pheromone deposited by ant k, i.e, the pheromone deposited becomes greater if the tour length is reduced (Dorigo & Stutzle, 2006).

The ant system algorithm for the travelling salesperson problem can be constructed for the given graph as:

G: (V, E) where V is the vertices, and E represents edges connecting.

1. Considering n as number of vertices, 0 and K as the number of ants to travel, where $K \leq n$, K can be placed at randomly selected positions.

2. Each of the ant complete the tour (Hamiltonian cycle) based on the following procedure.

3. The ant visits the vertices and stores its route in its memory, denoted as M_k^i for remembering nodes visited, where $M_k^i = \varnothing$ at initialization for all i, where the city visited is added to the memory.

4. An ant in city A can select any city B from the list of unvisited nodes by computing the probability, according to the equation. $\tau(x^a, y^b)^\alpha \cdot \gamma(x^a, y^b)^\beta / \Sigma_{j=1} \tau(x^a, Z)^\alpha \cdot \tau(x, z)^\beta$ is yet to be visited, or 0 in the value, where j is iteration for number of edges that are not present in the ants' memory M_k^i, α and β are constants. $\tau(x, y)$ is the pheromone update on edges, γ (a,b) is the inverse of the distance between a and b, and z is the member of the list of unvisited nodes.

5. The pheromone update here is calculated according to the formula:

$$\tau(a, b) \leftarrow (1 - \rho), \tau(x, y) + \sum_{i=1} \Delta \tau^i(x, y) \qquad (4.7)$$

where ρ (0,1] is an evaporation rate where i can range anywhere between 1 and m; $\Delta \tau^I$ (a,b) is amount of pheromone deposited on edge (a,b) by an ant i, which is computed by $\Delta \tau(a,b)$, the amount of pheromone deposited on edge (a,b) by a considered ant i, which is computed by $\Delta \tau(a,b) = c/L_i$ if (a,b) does not belong to the edges passed over by ant or contains a value 0. L_i = Length of journey traversed by ant I, and c is a constant.

6. Keep on looping until reaching the max iteration possible.

7. Return the length of most favorable tour.

The experiments yield the fact that the algorithm provided an optimum solution until a journey of 30 cities was reached. An increase in the number of cities to be visited beyond 30 led to the consumption of more time and poor or unfeasible solutions, making it unsuitable for use in solving the TSP when a large number of cities are invovled (Fogel et al., 1966; Jaillet, 1985, 1988).

4.8.2 Ant Colony System

The ant colony system is an modified version of the ant system designed to overcome the technical lacunas of the ant system by incorporating a local pheromone update to be done by ants in addition to the general pheromone update performed at the end of the construction process (Dorigo, 1992).

Here, the ants locally update the pheromone tables covering the last possible edge. The pheromone update is done using the formula:

$$\tau\,(x,y) \leftarrow (1-\rho) \cdot \tau\,(x,y) + \varphi \cdot \tau_0 \tag{4.8}$$

where ρ = Pheromone decay constant

τ_0 = Initial pheromone value

The global update rule is applied by the best performing ant, which provided the shortest rule since the commencement of the algorithm or at the end of the iteration using the formula

$\tau\,(x,\,y) \leftarrow (1-\rho) \cdot \tau\,(x,\,y) + \rho.\,L^{-1}{}_{best}$ where $\tau\,(x,\,y)$ remains same if $(x,\,y)$ is not contributing to the best ant's tour.

4.8.3 Max-Min Ant System

The Max-Min ant system (MMAS) can also be said to be a variation on the ant system that is a peculiarity characterized by the presence of a greedy search method. The algorithm explores the search history for the best possible solution and allows the increase of pheromone values only on these results (Dorigo & Stutzle, 2006). The experimental evaluation of Stutzle and Hoose in their research, as claimed by them, states that MMAS can be considered to be one of the best performing ACO algorithms for solving NP-Hard problems, like the TSP or QAP.

Here, the pheromone update can be calculated using eq. (4.8)

$$Z \leftarrow \tau(x,y) \leftarrow (1-\rho) \cdot \tau\,(x,y) + \Delta\tau^{best}\,(x,y) \tag{4.9}$$

When the pheromone is bounded by both the upper and the lower bound (τ_{min} & τ_{max}), where z will evaluate to any of the three values.

τ_{max} if $Z > \tau_{max}$, τ_{min} (lower bound) if $Z < \tau_{min}$ and $\Delta\tau^{best}$ (x, y) can be calculated as

$$\Delta\tau^{best}(x,y) \leftarrow \frac{1}{L_{best}} \tag{4.10}$$

4.8.4 Rank-Based Ant System

This system, also called ant system with elitist strategy, is a variant of the ACO algorithm that provides an improved result as compared to the ant system. The elitist strategy emphasizes the best solution constructed, which provides guidance to the successive solutions. Here, the collection of the fittest solutions is the hallmark of the elitist strategy, which provides more concentration over the local search. The AS_{Rank}, i.e., the ant-based method (Camazine et al., 2001) explores the performance gain of $AS_{elitist}$ strategy over the traditional AS algorithm to further improve the computational performance (Bullnheimer, Hartl, & Strauss, 1997, 1999).

Here, the pheromone update in AS_{Rank} algorithm can be done by:

$$\tau(x,y) \leftarrow (1-\rho) \cdot \tau(x,y) + \sum_{\mu=1} \Delta\tau^{\mu}(x,y) + \Delta\tau^{best}(x,y) \tag{4.11}$$

$$\Delta\tau^{\mu}(x,y) \leftarrow (\delta-\mu)Q/L\mu \tag{4.12}$$

$$\Delta\tau^{best}(x,y) \leftarrow \delta(Q/L_{best}) \tag{4.13}$$

where
δ = Number of elitist ants
μ = Ranking index
$\Delta\tau^{\mu}$ (x, y) = Increase in the trail concentration on the edge (i,j) used by the μth best ant in its tour.
$\Delta\tau^{best}$ (x, y) = Quantity of pheromone deposited on edge (i,j) used by elitist ant in its tour.
$L\mu$ = Tour length completed by μ^{th} best ant
L_{best} = Length of tour completed by the best ant.

4.8.5 Continuous Interacting ACO

The CIAC algorithm is specially designed to solve continuous problems unlike the discrete optimization problems solved by previous algorithms. Here, the concept of dense hierarchy is used where the ant colonies handle the information.

The CIAC Algorithms are composed of the following three steps:

Step 1. Parameters are initialized.

Step 2. The ant ranges are distributed over agent's population and positioned randomly within the search area.

Step 3. The algorithm is characterized by ants moving until the stop condition is met.

4.8.6 Continuous Orthogonal ACO

The ACO algorithm designed to solve the continuous problems uses the orthogonal-designed method for task accomplishment. The pheromone deposit strategy makes that ants capable of collaborative search for a solution. This algorithm is composed of

1. Orthogonal exploration
2. Global modulation

Orthogonal exploration – In orthogonal exploration, the ants choose the region for exploration based upon the pheromone values. The probability factor χ_0. defined by the user influences whether the ant chooses the region holding the highest pheromone value or uses the route wheel, selecting from a set of regions S_R.

The probability of choosing a region by an ant can be written as

$$P(i) = \tau_I | \sum_{K \in} S_R \tau_K \qquad (4.14)$$

τ_I = Pheromone in region i

K = A region in the set of region S_R

τ_k = Pheromone in region k

Steps undertaken by orthogonal explorations are:

Step 1. The route with highest pheromone deposit update is chosen.

Step 2. A unique orthogonal array is created by random selection of n unique columns of the given orthogonal array.

Step 3. N close points are selected.

Step 4. The radius of the considered region is adjusted.

Step 5. The center of the considered region is moved to the best point.

Global Modulation: After all the considered ants (m) are finished with orthogonal exploration, the evaluation of the desirability factor for every region is calculated.

A parameter $\Psi \in [0,1]$, i.e., the elitist rate, is utilized for selection of best regions in S_R and the best regions are ranked beginning from one, and the selected region is included in S_R, i.e., the set consisting of highly ranked regions.

The pheromone τ_I on the selected region i is changed by

$$\tau_I \leftarrow (1-\rho)\tau_I + \rho.\tau_0(v.\mu - \text{rank}_i + \text{visit}_i + 1) \qquad (4.15)$$

where
   $\rho \in (0, 1]$ is the rate of evaporation.
   τ_0 = Initial pheromone value
   μ = Number of regions
   rank_i = The rank of region i
   visit_i = Number of ants that visited region i
From the discussion in this section, we can segregate the collection of designed ACO algorithms into two categories that can solve both discrete and continuous optimization problems. The algorithms like the AS, AS_{rank}, ACS, MMAS are designed to handle the discrete optimization problems, whereas algorithms like CIAC and COAC are designed for problems belonging to the continuous domain. One of the important classification criteria of most ACO algorithms is the way pheromones are updated, which can be done either using

1. The ant system update rule
2. Iteration-based update rule
3. Best so far update rule

4.9 ACO for NP Hard Problems

NP hard problems can be defined as a "decision problem" H can be said to be NP hard when for every problem L belonging to NP; there exists a polynomial time reduction from L to H or the problem L can be solved in polynomial time if a subroutine call takes only single step computation.
   The majority of the problems belong to one of the following categories:

1. Routing
2. Assignment
3. Scheduling
4. Subset problems

4.9.1 Routing Problems

In routing problems, the ACO can be beneficial in solving sequential ordering problems as well as vehicle routing problems. The sequential ordering problem (SOP) focuses on looking for minimum weight Hamiltonian path on a directed graph, where arcs carry weight and nodes are subjected to precedence constraints. The SOP can be compared as an abstraction of the asymmetric TSP described as (Wang & Mendel, 1992):

- A solution connecting the beginning node to the final node using Hamiltonian path.
- Arcs carrying weight – the weights C_{ij} for arc (i,j) can be redefined by removing node weights from the original definition and adding a node that weighs P_j for node j for each arc incident to j that resulted in a new arc weighing $c_{ij} = c_{ij} + p_j$.
- Precedence constraints are defined among the nodes. If a node j precedes node i, it can be represented by weight $c_{i\,j} = -1$ for the arc from (i,j). Hence, if $c_{ij} > 0$, the weight gives cost of arc (i,j) where as $c_{ij} = -1$, may not be immediately but precedes node i.

The application of ACO to SOP is an modification of the ACO algorithms for the TSP, called a hybrid AS-SOP (HAS-SOP) (Gambardella & Dorigo, 2000). The HAS-SOP algorithm consists of variables defined as:

i. **Construction Graph**: A set of components containing a set of n nodes including the start node (N_o) and final node (N_{n+1}). Here the components are completely connected to each other where the solution Hamiltonian paths starts at node 0 ending at $n + 1$, complying with all precedence constraints.

ii. **Constraints**: The algorithmic constraints in SOP demand all nodes in the graph be visited just once and all precedence conditions be satisfied.

iii. **Pheromone Trail**: τ_{ij} represents the destiny to select node j while the node is at node i.

iv. **Heuristic information**: The HAS-SOP heuristic information gives $\eta_{ij} = \frac{1}{C'_{ij}}$ when $C'_{ij} \neq -1$, and $n_{ij} = 0$

v. **Solution Construct**: The ants start from node N_0, building Hamiltonian paths connecting node N_0 to N_{n+1}. The ants build solutions by choosing probabilistically the next node from the collection of feasible neighbours. For example, ant k at node i chooses node j, $j \in N_i^k$, with probability calculated by

$$J = \begin{cases} \underset{C \in N_i^k}{arg\,max}\{\tau_{il}[\eta_{il}]\} & \text{if } q \leq q_0 \\ J \text{ otherwise} \end{cases}$$

(4.16)

Where q is a random variable uniformly distributed in $[0,1]$, $q_0(0 \leq q_0 \leq 1)$ and J is a random variable according to probability distribution with $\alpha = 1$, where α_i^k is a set of all unvisited nodes j.

Pheromone Update: The algorithms use a common pheromone updating procedure such as that used in TSP or ACS.

Local Search: The local search is a three opt or SOP-3-exchange (Gambardella & Dorigo, 2000) procedure that can handle smoothly the multiple precedence constraints without effecting computational complexity.

Results: The results obtained with HAS-SOP are extremely positive compared to state-of-the-art algorithms for SOP, genetic algorithm being outstanding in performance compared to the rest of HAS-SOP .

4.9.2 Vehicle Routing

The vehicle routing problem (VRP) is considered as one of the main problems in distribution management (Toth & Vigo, 2001). In the capacitated VRP/CVRP, one central department has to cater n customers, identified as index 0. Each customer I has a demand (positive) belonging to same merchandise, where d_{ij} is the travel time given between a pair of customers i and j, (i, j).

The problem aims to find a route set that can possibly reduce the total travel time such that

1. A customer is served exactly once by a little vehicle.
2. The vehicle route starts as well as ends at a depot.
3. The total demand reached should not be more than its capacity B.

The CVRP is supposed to be an NP hard problem, since it contains TSP as its sub-problem.

4.9.3 Assignment Problem

Here, the problem concentrates on assigning a set of objects to a set of resources with respect to certain constraints.

The assignment can be depicted as a mapping of objects from set I to set J, where the objective function to minimize is a function of the assignment done (Lourenço & Serra, 1998).

The solution starts with the mapping of the problem on construction graph $G_C = (C, L)$, where
C = Set of components
L = Set of arcs connecting the graph
The construction process allows ants to walk on a construction graph.

The assignment problem can be of various kinds such as quadratic, generalized, frequency, etc.

4.9.4 Scheduling Problem

There have been a variety of scheduling problems undertaken by ACO, where different ACO variants are used for problems with different natures. The construction graph here is shown by a set of jobs or operations to be assigned to the machine to operate upon. In certain cases, the ordering of jobs varies the results obtained to the extent that job ordering becomes a crucial factor (Grosso, Della Croce, & Tadei, 2004).

4.10 Current Trends in ACO

ACO is a very prominent and up-and-coming field in which to work within areas of current research. In a few of these research fields, such as dynamic problems, stochastic problems, and multiple-objective problems, ACO has been proving a very efficient technique to achieve solutions.

Dynamic problems have instance data values that keep changing; objective function values, decision parameters, or constraints might be replaced while solving problems. Whereas stochastic problems are those that have single probabilistic information regarding the objective function values, decision variable values or constraint boundaries, due to uncertainty, noise, approximation, or other factors. Multiple-objective problems are those where a multi-objective function is used to access the solution quality of competing criteria. We discuss the details of these problems as:

1. **Stochastic Optimization Problems**: In these types of problems, a few variables involved are stochastic in nature. It can be a problem component, part of a problem, or values adopted by few variables defining the problem or value returned by the objective function. Typical examples of ACO applied to stochastic optimization problems are network routing or probabilistic TSP.

2. **Multi-Objective Optimization Problems (MOOPs)**: Many problems in real-world applications require the evaluation of multiple as well as conflicting objectives. In these kinds of problems, the aim is finding a solution that gives the best compromise among various objectives. A solution can be termed as efficient if it is not dominated by other solutions, and a Pareto optimal set can be defined as a set consisting of all efficient solutions.

3. **Parallelization:** The solutions to real-world problems can involve a huge number of computations involving a great amount of time and

complexity, even when using metaheuristics. Hence, parallel imple-mentations of ACO Aagorithms for running on distributed (parallel) computing hardware are desirable.

The ant colony optimization mechanism is as such a distributed concept, where multiple small artificial ants work together, achieving a common goal. Hence, it can be said to provide parallelization through a number of parallel versions (run-time reconfiguration) of ACO that have been implemented; it still challenges researchers to conceive an efficient parallel version that can provide a considerable performance gain over the general ACO.

4.11 Application of ACO in Different Fields

ACO algorithms are not only applicable to research problems in the field of computer science but are also useful in various other streams of science and technology. The application areas of ACO range over a variety of

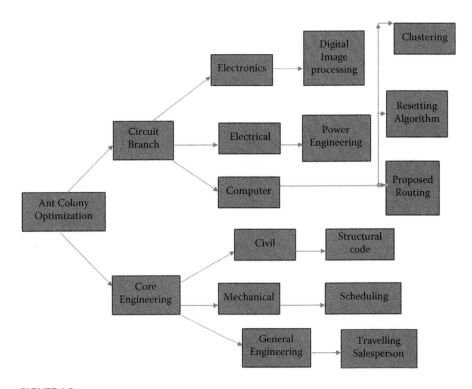

FIGURE 4.5
Application areas of ant colony optimization (Chandra Mohan & Baskaran, 2012).

combinatorial optimization problems, as shown in Figure 4.5 (Chandra Mohan & Baskaran, 2012).

We have already discussed the role of ACO in solving the TSP in Section 4.6. Therefore, in this section, we will focus on other contributions of ACO in the various fields as mentioned in the Figure 4.5.

4.11.1 Scheduling Problems

One of the challenging combinatorial problems of this current era is the job scheduling problem, which evolved out of the rapid growth in demand (Szwarc, Grosso, & Della Croce, 2001).

The job scheduling problem can be defined as:

1. *N* is the number of independent jobs requiring to be scheduled.
2. *M* is the number of machines where jobs have to be assigned.
3. *W* is the workload of each job.
4. *C* is the computing capacity of each machine.
5. The ready time (*T*) denotes that all previous jobs are finished with execution.
6. *ETC* is the expected execution time of job *i* in machine *j*.

The target is to schedule the jobs to the machines so as to get the highest performance output.

Many approaches have been adopted and successfully applied to this problem, such as simulated annealing, tabu search, genetic algorithms, etc. A few researchers have proposed an integrated evolutionary algorithm with some optimization strategies. Xing Chen, Wang Zhao, and Xiong in 2010 designed a framework: the Knowledge-Based Heuristic Searching Architecture (KBHSA), which integrates the knowledge model and the heuristic searching model to look for optimal solutions. This KBACO architecture, when applied to common benchmark problems, outperformed the previous approaches while solving Flexible Job Shop Scheduling Problem (FJSSP).

4.11.2 Structural and Concrete Engineering

Structural optimization is an extremely appealing field of research in the area of combinatorial optimizations (discrete and continuous optimization). Stochastic search methods that mimic the biological methods or natural phenomena have been sought after to handle these kinds of problems.

The structural optimization problem deals with designing of steel trusses that requires selecting members from standard steel pipe, section table such that it can satisfy the strength and serviceability requirement specified by selected code of practice while maintaining the economy factor or overall material cost of the structure (Martínez, González-Vidosa, Hospitaler, & Yepes, 2010).

Martínez et al. (2010) have proposed an ACO model that deals with the economic optimization of the reinforced concrete (RC) bridge piers having hollow rectangular sections.

4.11.3 Digital Image Processing

As we know, ACO is an optimization technique that can be used for a variety of applications dealing with either continuous or discrete optimization problems, ACO hence has made a successful contribution to digital image processing in solving problems related to images such as image edge detection, image compression, image segmentation, structural damage monitoring, etc. (Grosso et al., 2004).

The following are a few of the research contributions in this area. Yu, Au, Zou, Yu, and Tian (2010) proposed ant colony – fuzzy C-means hybrid algorithms (AFHA) for adaptively clustering the image pixels that are viewed as three-dimensional data pieces (You, Liu, & Wang, 2010). Tan and Isa (2011) have contributed a novel histogram threshold – fuzzy C-means hybrid approach (FTHCM) that is capable of finding different patterns in applications in pattern recognition as well as computer vision, specifically in colour image segmentation.

In electrical engineering, the problem of economic dispatch is a vital issue in areas of mathematical optimization regarding the power system operation that needs immediate attention. The rapid rise in the population has generated increased demand on thermal power plants. The problem of environmental or economic dispatch has attracted much concern regarding the design of a good dispatch scheme that can produce a good economic return as well as reduce the emission of pollutants.

This issue was addressed efficiently by using the ant colony optimization method.

Cai, Ma, Li, Li, and Peng (2010) have proposed a multi-objective chaotic ant swarm optimization (MOCASO) that can efficiently address the problem of balancing economic growth and the reduction of pollution. Pothiya, Ngamroo, and Kongprawechnon (2010) have also contributed an ACO model for the economic dispatch problem with non-smooth cost functions with valve-point effects and multiple fuels.

4.11.4 Clustering

Clustinger, which may also be denoted as a set partitioning problem, is a universally applied method that covers mathematical programming, graph theory, scheduling and assignment, network partitioning, routing, etc. The main objective of clustering is to group the objects into sets of mutually exclusive clusters in order to achieve the minimum or maximum of an objective function. There are many proposed clustering techniques, such as hierarchical clustering, partition-based clustering, and artificial-intelligence-based

K means, currently the most popular technique adopted to cluster large databases, but they still suffer shortcomings such as (Verbeke, Martens, Mues, & Baesens, 2011).

1. They are dependent on initial state and convergence to local optima or
2. Global solutions of large problems cannot be obtained with reasonable amount of computation effort

Hence researchers have tried ACO to come up with solutions that can overcome these issues.

4.11.5 Routing Problem

One of the major contributions of ACO to the field of networking is addressing routing issues. A few approaches, such as ant-net routing using ACO techniques are valuable contributions that have provided better results in comparison to others due to their real-time computation and low control overhead. New technology in the field of networking and wireless technology, where the mode of communication is moving toward concepts like ad hoc, sensors, wireless maintaining a resilient robust communication medium, and fast communication is becoming all the more challenging. The ACO technique is one of those very much in demand and has been adopted by the researchers to address these networking problems.

4.12 Conclusion

The chapter focuses mainly on the concepts of ant colony optimization and how it can be implemented in solving various discrete as well as continuous optimization problems. We further discuss its application areas, covering how ACO has provided promising solutions in divergent fields and thus enjoys the position of being the most important and sought-after technique in the whole of swarm intelligence family.

References

Blum, C., & Dorigo, M. (2004). The hyper-cube framework for ant colony optimization. *IEEE Transactions on Systems, Man, and Cybernetics, Part B (Cybernetics), 34* (2), 1161–1172.
Blum, C., Roli, A., & Dorigo, M. (2001). HC–ACO: The hyper-cube framework for ant colony optimization. In *Proceedings of MIC'2001—Metaheuristics International Conference* (Vol. 2, pp. 399–403). Porto, Portugal, July 16–20, 2001.

Bonabeau, E., Dorigo, M., & Theraulaz, G. (1999). *Swarm Intelligence: From Natural to Artificial Systems*. New York: Oxford University Press.

Bonabeau, E., Dorigo, M., & Theraulaz, G. (2000). Inspiration for optimization from social insect behavior. *Nature, 406,* 39–42.

Bonabeau, E., Sobkowski, A., Theraulaz, G., & Deneubourg, J.-L. (1997). Adaptive task allocation inspired by a model of division of labor in social insects. In D. Lundha, B. Olsson, & A. Narayanan (Eds.), *Bio-Computation and Emergent Computing* (pp. 36–45). Singapore: World Scientific Publishing.

Bullnheimer, B., Hartl, R. F., & Strauss, C. (1997). *A New Rank-Based Version of the Ant System—A Computational Study*. Technical report. Austria: Institute of Management Science, University of Vienna, Austria.

Bullnheimer, B., Hartl, R. F., & Strauss, C. (1999). Applying the ant system to the vehicle routing problem. In S. Voss, S. Martello, I. H. Osman, & C. Roucairol (Eds.), *Meta-heuristics: Advances and Trends in Local Search Paradigms for Optimization* (pp. 285–296). Dordrecht, The Netherlands: Kluwer Academic Publishers.

Cai, J., Ma, X., Li, Q., Li, L., & Peng, H. (2010). A multi-objective chaotic ant swarm optimization for environmental/economic dispatch. *International Journal of Electrical Power & Energy Systems, 32*(5), 337–344.

Camazine, S., Deneubourg, J. L., Franks, N. R., Sneyd, J., Bonabeau, E., & Theraula, G. (2001). *Self-organization in Biological Systems* (Vol. 7). Princeton University Press.

Chandra Mohan, B., & Baskaran, R. (2012). A survey: Ant colony optimization based recent research and implementation on several engineering domains. *Expert Systems with Application, 39,* 4618–4627.

Deneubourg, J.-L., Aron, S., Goss, S., & Pasteels, J.-M. (1990). The self-organizing exploratory pattern of the Argentine ant. *Journal of Insect Behavior, 3,* 159–168.

Deneubourg, J.-L., Goss, S., Franks, N., Sendova-Franks, A., Detrain, C., & Chrétien, L. (1991). The dynamics of collective sorting: Robot-like ants and ant-like robots. In J.-A. Meyer & S. W. Wilson (Eds.), *Proceedings of the First International Conference on Simulation of Adaptive Behavior: From Animals to Animats* (pp. 356–363). Cambridge, MA: MIT Press.

Dorigo, M. (1992). *Optimization, Learning and Natural Algorithms* (PhD thesis). Dipartimento di Elettronica, Politecnico di Milano, Milan [in Italian].

Dorigo, M., & Di Caro, G. (1999). Ant colony optimization: A new metaheuristic. In *Evolutionary Computation, 1999. CEC 99. Proceedings of the 1999 Congress on* (Vol. 2, pp. 1470–1477). IEEE.

Dorigo, M., Di Caro, G., & Gambardella, L. M. (1999). Ant algorithms for discrete optimization. *Artificial Life, 5*(2), 137–172.

Dorigo, M., Maniezzo, V., & Colorni, A. (1996). Ant system: Optimization by a colony of cooperating agents. *IEEE Transactions on Systems, Man, and Cybernetics, Part B (Cybernetics), 26*(1), 29–41.

Fogel, L. J., Owens, A. J., & Walsh, M. J. (1966). Artificial intelligence through simulated evolution. From chaos to order: How ants optimize food search. Retrieved from https://www.pik-potsdam.de/news/press-releases/archive/2014/from-chaos-to-order-how-ants-optimize-food-search (Accessed on 25 February 2018).

Glover, F., & Kochenberger, G. (Eds.). (2002). *Handbook of Metaheuristics*. Norwell, MA: Kluwer Academic Publishers.

Goldberg, D. E. (1989). *Genetic Algorithms in Search, Optimization and Machine Learning*. Reading, MA: Addison-Wesley.

Goss, S., Aron, S., Deneubourg, J. L., & Pasteels, J. M. (1989). Self-organized shortcuts in the Argentine ant. *Naturwissenschaften*, 76(12), 579–581.

Grosso, A., Della Croce, F., & Tadei, R. (2004). An enhanced dyna search neighbourhood for the single-machine total weighted tardiness scheduling problem. *Operations Research Letters*, 32(1), 68–72.

Handl, J., & Meyer, B. (2002). Improved ant-based clustering and sorting in a document retrieval interface. In J. J. Merelo, P. Adamidis, H.-G. Beyer, J.-L. Fernandez-Villacanas, & H.-P. Schwefel (Eds.), *Proceedings of PPSN-VII, Seventh International Conference on Parallel Problem Solving from Nature, Vol. 2439 in Lecture Notes in Computer Science* (pp. 913–923). Berlin: Springer-Verlag.

Holland, J. (1975). *Adaptation in Natural and Artificial Systems*. Ann Arbor: University of Michigan Press.

Holland, O., & Melhuish, C. (1999). Stigmergy, self-organization, and sorting in collective robotics. *Artificial Life*, 5(2), 173–202.

Interesting facts about ants. http://justfunfacts.com/interesting-facts-about-ants/ (Accessed on 25 February 2018).

Jacobs, L. W., & Brusco, M. J. (1995). A local search heuristic for large set covering problems. *Naval Research Logistics*, 42, 1129–1140.

Jaillet, P. (1985). *Probabilistic Traveling Salesman Problems* (PhD thesis). Cambridge, MA: MIT.

Jaillet, P. (1988). A priori solution of a travelling salesman problem in which a random subset of the customers are visited. *Operations Research*, 36(6), 929–936.

Johnson, D. S., & McGeoch, L. A. (1997). The travelling salesman problem: A case study in local optimization. In E. H. L. Aarts & J. K. Lenstra (Eds.), *Local Search in Combinatorial Optimization* (pp. 215–310). Chichester, UK: John Wiley & Sons.

Kan, A. R., & Shmoys, D. B. (1985). *The Traveling Salesman Problem: A Guided Tour of Combinatorial Optimization* (Vol. 3, pp. 1–463). E. L. Lawler & J. K. Lenstra (Eds.). New York: Wiley.

Knox, J. (1994). Tabu search performance on the symmetric travelling salesman problem. *Computers & Operations Research*, 21(8), 867–876.

Kube, C. R., & Zhang, H. (1994). Collective robotics: From social insects to robots. *Adaptive Behavior*, 2, 189–218.

Lourenço, H. R., Martin, O. C., & Stützle, T. (2002). Iterated local search. In *Handbook of metaheuristics* (pp. 320–353). Boston, MA: Springer.

Lourenço, H. R., & Serra, D. (1998). *Adaptive Approach Heuristics for the Generalized Assignment Problem*. Technical report no. 304. Barcelona, Spain: Department of Economics and Management, Universitat Pompeu Fabra.

Lumer, E., & Faieta, B. (1994). Diversity and adaptation in populations of clustering ants. In J.-A. Meyer & S. W. Wilson (Eds.), *Proceedings of the Third International Conference on Simulation of Adaptive Behavior: From Animals to Animats* (pp. 501–508). Cambridge, MA: MIT Press.

López-Ibáñez, M., Stützle, T., & Dorigo, M. (2016). Ant colony optimization: A component-wise overview. *Handbook of Heuristics*, 1–37.

Maniezzo, V., & Milandri, M. (2002). An ant-based framework for very strongly constrained problems. In M. Dorigo, G. Di Caro, & M. Sampels (Eds.), *Proceedings of ANTS 2002—From Ant Colonies to Artificial ANTS: Third International*

Workshop on Ant Algorithms, Vol. 2463 of Lecture Notes in Computer Science (pp. 222–227). Berlin: Springer-Verlag.

Martínez, F. J., González-Vidosa, F., Hospitaler, A., & Yepes, V. (2010). Heuristic optimization of RC bridge piers with rectangular hollow sections. *Computers & Structures, 88*(5–6), 375-386.

Martinoli, A., & Mondada, F. (1998). Probabilistic modelling of a bio-inspired collective experiment with real robots. In T. L. R. Dillman, P. Dario, & H. Wörn (Eds.), *Proceedings of the Fourth International Symposium on Distributed Autonomous Robotic Systems (DARS-98)* (pp. 289–308). Berlin: Springer-Verlag.

Papadimitriou, C. H., & Steiglitz, K. (1982). *Combinatorial Optimization—Algorithms and Complexity.* Englewood Cliffs, NJ: Prentice Hall.

Pothiya, S., Ngamroo, I., & Kongprawechnon, W. (2010). Ant colony optimisation for economic dispatch problem with non-smooth cost functions. *International Journal of Electrical Power & Energy Systems, 32*(5), 478–487.

Rechenberg, I. (1973). *Evolutionsstrategie—Optimierung Technischer Systeme Nach Prinzipien Der Biologischen Information.* Freiburg, Germany: Fromman Verlag.

Reinelt, G. (1994). *The Traveling Salesman: Computational Solutions for TSP Applications, Vol. 840 of Lecture Notes in Computer Science.* Berlin: Springer-Verlag.

Robinson, G. E. (1992). Regulation of division of labor in insect societies. *Annual Review of Entomology, 37,* 637–665.

Romeo, F., & Sangiovanni-Vincentelli, A. (1991). A theoretical framework for simulated annealing. *Algorithmica, 6*(3), 302–345.

Schwefel, H. P. (1981). *Numerical Optimization of Computer Models.* Chichester, UK: John Wiley & Sons.

Stutzle, T., & Dorigo, M. (2002). A short convergence proof for a class of ant colony optimization algorithms. *IEEE Transactions on Evolutionary Computation, 6*(4), 358–365.

Szwarc, W., Grosso, A., & Della Croce, F. (2001). Algorithmic paradoxes of the single machine total tardiness problem. *Journal of Scheduling, 4*(2), 93–104.

Toth, P., & Vigo, D. (Eds.). (2001). *The Vehicle Routing Problem. SIAM Monographs on Discrete Mathematics and Applications.* Philadelphia: Society for Industrial & Applied Mathematics.

Verbeke, W., Martens, D., Mues, C., & Baesens, B. (2011). Building comprehensible customer churn prediction models with advanced rule induction techniques. *Expert Systems with Applications, 38,* 2354–2364.

Wagner, I. A., Lindenbaum, M., & Bruckstein, A. M. (1996). Smell as a computational resource—A lesson we can learn from the ant. In M. Y. Vardi (Ed.), *Proceedings of the Fourth Israeli Symposium on Theory of Computing and Systems (ISTCS-99)* (pp. 219–230). Los Alamitos, CA: IEEE Computer Society Press.

Wagner, I. A., Lindenbaum, M., & Bruckstein, A. M. (1998). Efficient graph search by a smell-oriented vertex process. *Annals of Mathematics and Artificial Intelligence, 24,* 211–223.

Wagner, I. A., Lindenbaum, M., & Bruckstein, A. M. (2000). ANTS: Agents, networks, trees and sub-graphs. *Future Generation Computer Systems, 16*(8), 915–926.

Wang, L. X., & Mendel, J. M. (1992). Generating fuzzy rules by learning from examples. *IEEE Transactions on Systems Man, and Cybernetics, 22*(6), 1414–1427.

Xing, L. N., Chen, Y. W., Wang, P., Zhao, Q. S., & Xiong, J. (2010). A knowledge-based ant colony optimization for flexible job shop scheduling problems. *Applied Soft Computing, 10*(3), 888–896.

You, X.-M., Liu, S., & Wang, Y.-M. (2010). Quantum dynamic mechanism-based parallel ant colony optimization algorithm. *International Journal of Computational Intelligence Systems, 3*(Suppl. 1), 101–113.

5

Particle Swarm Optimization

Shanthi M. B., D. Komagal Meenakshi, and Prem Kumar Ramesh

CONTENTS

5.1 Particle Swarm Optimization – Basic Concepts

Particle swarm optimization (PSO) is a population-based metaheuristics inspired by nature. Dr. Russell C. Eberhart and Dr. James Kennedy introduced it in 1995 (Kennedy & Eberhart, 1995).

5.1.1 Introduction: Background and Origin

The particle swarm concept originated with the effort of Reeves (1983), who came up with the concept of particle systems to represent dynamic objects that cannot be easily illustrated by surface-based models. He developed a system of particles considered as independent entities that work in harmony to imitate a fuzzy object. Reynolds (1987) then added a concept of communication between the particles, social behaviour, with the help of a flocking algorithm, whereby each particle adheres to the flocking rules. The ideas from the dynamic theory of social impact by Nowak, Szamrej, and Latané (1990) also helped in understanding the principles underlying how individuals are affected by the social environment. In addition to this, Heppner and Grenander (1990) related a *roost* concept, i.e., the flock aims for some roosting area. In these systems, the particles are autonomous, but their movements are regulated by a class of rules. These observations on collective behaviours in these social animals led toward the implementation of this model in solving different optimization problems. Initially, Eberhart and Kennedy wanted to simulate the unforeseeable choreography of birds but then later realized the potential optimization of a flock of birds.

5.1.2 Features of PSO Technique

The PSO technique encompasses the following features. PSO is a metaheuristic because it makes almost nill or very few inferences about the optimization problem and it has the ability to search very large space with distributed candidate solutions. PSO exhibits swarm intelligence in its optimization process. It mainly follows five basic principles observed in SI based algorithms. Mark Millonas (1994) (Millonas, 1993) has stated these principles are followed by the particles while communicating with other fellow particles in the swarm. The initial one is the proximity principle, which has been highlighted in time- and space-bounded computations.

The second principle states that the particles of the swarm should be able to react based on the quality factors of the environment. Third principle has a focus on diversification in the search process. Fourth is the principle that states the particles in the swarm have to maintain stability even when there is a change in the environment. The fifth principle states that the population must be able to modulate its mode of behaviour when it is worth the price of computation. It relies on self-organization, which is a key trait of SI. The concept is very simple and hence the algorithm consumes neither time nor memory. PSO does not demand the optimization problem to be differentiable because it neglects the gradient of the optimization problem.

5.1.3 Similarities and Differences between PSO, EA, and GA

PSO, evolutionary algorithms (EA), and genetic algorithms (GA) initiate with a group of randomly created population members. In PSO, a particle is comparable to a population member (chromosome) in a GA and EA. Both evaluate the population based on a fitness function. The main difference between the PSO approach as compared to EA and GA is that PSO does not involve genetic operators such as crossover and mutation operators (Eberhart & Shi, 1998). The inter-particle communication mechanism in PSO is remarkably different from that in EA. In EA approaches (Jones, 2005), chromosomes carry the information to the next generation based on effective cultural habits. In PSO, only the 'best' particle renders the information to others. It is a unidirectional information sharing mechanism; the evolution only looks for the best solution. Compared to GAs, PSO is simple and uncomplicated to implement as it has only few parameters to adjust.

5.1.4 Overview of Particle Swarms

The particle swarm is not merely a collection of particles. A particle alone is powerless to solve any kind of problem; headway occurs only when the particles communicate. Each member of the swarm interacts with others using simple and limited actions. Particle swarms usually use three components to determine the searching behaviour of each individual: a social component, a cognitive component, and an inertia component. The cognitive component is responsible for weighing the importance an individual gives to its own knowledge of the world. The social component weighs the importance an individual with respect to the cumulative knowledge of the swarm as a whole. Finally, the inertia component specifies how fast individuals move and change direction over time. In PSO, each particle in the swarm travels through the multidimensional search space and updates its position in subsequent iterations with its own knowledge and that of the fellow particles to achieve an optimal solution. Therefore, the PSO algorithm is a member of swarm intelligence.

5.2 PSO Variants

Due to its powerful capability and relatively low number of parameters, PSO has drawn wide attention since its inception. However, the original version of PSO had some deficiencies such as premature convergence and failed in finding optimal solutions for multimodal problems. One way to strengthen the capability of PSO is to dynamically adapt its parameters when running the particles' evolution process. Better results in PSO have been realized through different augmentations of the generic method, applying parameter tuning, changing topology, introducing intelligent randomized search strategies, and hybridizing it with other classical optimization techniques. The major variations in PSO are discussed in this chapter.

5.2.1 Unified PSO

Unified PSO (UPSO) is the combined version of local PSO and the global PSO. It was introduced by Parsopoulos and Vrahatis (2004b). It unifies the exploitation and exploration processes in a bounded space. The global version of PSO exhibits exploration, whereas the local version exhibits exploitation. During exploration, the algorithm tries to explore unknown candidate solutions distributed in the global search space, and during exploitation the search algorithm brings forth the solution found in the vicinity within a local neighbourhood. In the global version, all the particles move toward a single particle identified as global best, thereby increasing the convergence speed. It has better exploration ability in the global neighbourhood. In the local variant (Eberhart & Kennedy, 1995), each particle shares its best-ever position to that time, with other particles in the local neighbourhood. Hence, attraction toward the best particle is low, and the swarm would get trapped in local optima. Therefore, selection of neighbourhood size plays a major role in balancing the exploration and exploitation. The unified scheme of PSO brings out the best features of global and local variants of PSO and gives better results in terms of faster convergence and increased performance.

Classical PSO updates the particles' velocity based on the approach of global and local search. The global variant of PSO updates the velocity vector in the global search space as given in Equation (1), and the local variant updates as given in Equation (2).

$$G_i^{t+1} = \chi \left[v_i^t + c_1 r_1 \left(p_i^t - x_i^t \right) + c_2 r_2 \left(g_i^t - x_i^t \right) \right] \tag{1}$$

$$L_i^{t+1} = \chi \left[v_i^t + c_1 r_1 \left(p_i^t - x_i^t \right) \right] \tag{2}$$

where the term x_i denotes the current location of the particle, v_i denotes the particle, p_i denotes the particle best, χ denotes the extinction factor to avoid swarm explosion, and g_i denotes the global best position of the particle.

UPSO combines the two search vectors given in Equation (1) and in Equation (2) by adding a unification factor U to get the unified velocity vector as given in Equation (3)).

$$U_i^{t+1} = UG_i^{t+1} + (1-U)L_i^{t+1} \tag{3}$$

where, U takes the values in [0, 1]. For $U = 1$ in Equation (3), UPSO works as a global variant and when $U = 0$, it acts as a local variant of PSO. For all the values of U [0, 1] defines selection of range of values for exploration and exploitation in order to derive the best possible outcomes of unification.

5.2.2 Memetic PSO

Memetic particle swarm optimization (MPSO) is a modified version of PSO which combines PSO (Moscato, 1989), a popular global search technique, with memetic algorithm (MA) (Petalas, Parsopoulos, & Vrahatis, 2007), an adaptive local search algorithm. The main inspiration behind MA is Dawkins MEME (Dawkins, 1976), which mimics the natural behaviour of nature's system, which adopts the cultural changes over the period of time. MA is mainly used in multimodal optimization problems to derive multiple feasible solutions in parallel and leads toward the global solution when hybridized with PSO. Hybridization of PSO and MA is illustrated in the following section.

Particle velocity is updated as per Equation (4).

$$v_{i,j}^{t+1} = \chi \left[w v_{i,j}^t + c_1 r_{1,j} \left(p_{i,j}(t) - x_{i,j}(t) \right) + c_2 r_{2,j} \left(g_{i,j}(t) - x_{i,j}(t) \right) \right] \tag{4}$$

$$x_i^{t+1} = x_i^t + v_i^{t+1} \tag{5}$$

The term w is called 'inertia weight' and χ is extinction factor to control swarm explosion. The term c_1 is the 'cognitive parameter', and the term c_2 is the 'social parameter'. The terms r_1 and r_2 are the random vectors. The terms p_i and g_i are the local and global best locations in the swarm.

In global PSO, as all the particles tend to move toward a single global best particle by exploring the entire global neighbourhood. This leads to global exploration. It finds a single global solution. Multimodal problems work in two levels. In the first level, they aim to find multiple feasible solutions across the search space by dividing it into a number of local swarms of particles and finding the best solution for each of the local swarm. At the second level, all the local swarms will lead to finding the best solution among solutions from multiple particle swarms.

MPSO is a hybrid algorithm generated by combining a global PSO algorithm with a local MA algorithm. Based on the searching *strategies*, the algorithm can be constructed based on the following criteria.

- Apply local search to find best particle p_g in the swarm, where g denotes the index of the global best particle.
- For each particle's best position p_i, where $n = 1, 2, ..., N$, generate a random number r and, if its value is less than a predefined threshold value, apply local search on p_i.
- Apply local search on the particle p_g and a randomly selected p_i, where, $n = 1, 2, ..., N$ for which $||p_g - p_i|| > c \, \Delta \, (S)$, where c has values $[0, 1]$ and S is the diameter of the search space.
- Apply local search on the particle p_g and a randomly selected pi, where, $n = 1, 2, ..., N$.

The schema given above can be applied in every iteration of the algorithm or after a set of iterations. The Figure 5.1 illustrates the procedural steps followed in MPSO algorithm.

MPSO makes use of global variant of PSO to identify global optima in the global search space and MA to find the best optimal point in the local neighbourhood. Initially, the search algorithm initializes one or two population members and applies local search on them. It then applies evolutionary operators on the produced optimal solutions, to generate offspring. Local search is applied on these offspring, and a selection criterion is applied to select the best fit population as the parent population for the next generation. The process continues across generations until the stopping criteria are met.

5.2.3 VEPSO

Vector-evaluated particle swarm optimization (VEPSO) (Parsopoulos, Tasoulis, & Vrahatis, 2004) is a multi-objective PSO introduced by Parsopoulos and Vrahatis. It uses the concepts of multi-swarms from the vector-evaluated genetic algorithm (VEGA) (Schaffer, 1984). In VEGA, subpopulations are chosen based on each of the objectives, from the previous generation, separately. Then all these subpopulations are shuffled, and the new population is generated by applying crossover and mutation techniques. These ideas have been re-formed to fit the PSO framework. The VEPSO algorithm has been widely used to deal with multi-objective optimization (MOO) problems. MOO problems imply the simultaneous optimization of multiple objective

Input Search space S, N, c_1, c_2 and objective function F

Set Iteration counter $t = 0$

Initialize $x_i, v_i,$ and p_i

Evaluate $F(x_i)$ and determine indices g_i for $n = 1, 2, ..., N$

While (stopping criterion has not reached) Do

Updated velocity v_i for $n = 1, 2, ..., N$

Set $x_i^{t+1} = x_i^t + v_i^{t+1}$ for $n = 1, 2, ..., N$

Constrain each particle within S

Evaluate $F(x_i^{t+1})$ for $n = 1, 2, ..., N$

If $F(x_i^{t+1})) < F(x_i)$ then $p_i^{t+1} = x_i^{t+1}$

Else $p_i^{t+1} = p_i^t$t

Updated the indices g_i

While (local search is applied) Do

Apply local search and obtain the new solution y

End while

Set $t - t + 1$

End while

FIGURE 5.1
Memetic PSO algorithm.

functions, which generally conflict with one another. A single solution cannot satisfy all the objective functions due to the conflict between them. Therefore, a solution to the MOO problems usually is a set of trade offs or non-dominated solutions. The concept of VEPSO is based on multiple-objective functions, which are evaluated separately by multiple different swarms, but the information gained from other swarms influences its motion in the search space. The information exchange among swarms thus leads to Pareto optimal points. Then, with the help of co-evolutionary techniques, VEPSO searches for multiple feasible solutions in a larger solution space, under a single iteration. In co-evolutionary algorithms, the fitness of an individual in a swarm depends on the individuals of the other swarms. This improves the efficiency of the algorithm for better analysis and exploitation of the search space to find the set of non-dominant solutions. These solutions are used to detect the concave, convex, or partially convex or concave or discontinuous Pareto fronts (Coello Coello, & Sierra, 2004; Schaffer, 1984) more precisely. For multiple objective functions, each swarm works on one objective function using the information of *gBest* (*t*) gathered from the other swarm. The working steps of the algorithm are listed in the Figure 5.2.

Begin

   Initialize *position & velocity* for all M-swarm;

   Evaluate *objective* for all M-swarm;

   Initialize *archive*;

   Initialize *pBest* for all M-swarm;

   Initialize *gBest* for all M-swarm;

   While $i \le i$-max do

      Updated *velocity* for all M-swarm;

      Updated *position* for all M-swarm;

      Evaluate *objective* for all M-swarm;

      Updated *archive*;

      Updated *pBest* for all M-swarm;

      Updated *gBest* for all M-swarm;

      i++;

   End While

End

FIGURE 5.2
Algorithm for VEPSO.

In the VEPSO algorithm, the $pBest(t)$ which has the best fitness with respect to the mth objective, is the $gBest(t)$ for the mth swarm and is given in Equation (6)

$$gBest^m = \left\{ pBest^i \in S^m \mid f_m(pBest^i) = min\ f_m(\forall pBest^i \in S^m) \right\} \qquad (6)$$

On the other hand, the formula for velocity update is different and it is given in Equation (7). The particles in the mth swarm fly using $gBest^k(t)$ where k is given by Equation (8).

$$v_n^{mi}(t+1) = \chi \left[w v_n^{mi}(t) + c_1 r_1 \left(pBest_n^{mi} - P_n^{mi}(t) \right) \right.$$
$$\left. + c_1 r_1 \left(gBest_n^k - P_n^{mi}(t) \right) \right] \qquad (7)$$

$$k = \begin{cases} M, & m = 1 \\ m-1, & otherwise \end{cases} \qquad (8)$$

Once the objective functions are evaluated, the non-dominant solutions are recorded in an archive. Comparing each particle's fitness function with others, before comparing it with the non-dominant solutions in the archive, using the

Pareto optimality criterion leads to an archive with only non-dominant solutions. These non-dominant solutions are considered to be the possible solutions to the problem. VEPSO algorithm behaves in a similar manner to that of the PSO while handling MOO problems, but the difference is few processes are repeated for all *M*-swarm and non-dominant solutions are maintained in an archive.

5.2.4 Composite PSO (CPSO)

Parsopoulos and Vrahatis (2002) have introduced a new variant of PSO family called composite PSO (Parsopoulos & Vrahatis, 2004a). CPSO applies a novel learning strategy and assisted search mechanism. It employs an evolutionary algorithm called deferential evolution (DE), to control various parameters of PSO optimization in the learning stage. A combination of the historical information about the particle and the global best information is used to guide the particles to make the next move in the search space. CPSO maintains two particle populations to apply the meta-strategy of concurrent execution. Standard PSO swarm intends to minimize the functional value of the objective function, whereas the DE population of PSO parameter values reduces the convergence time by increasing the overall performance of the optimization process.

5.2.4.1 Differential Evolution (DE) Algorithm

Differential evolution (DE) is a population-based SI algorithm introduced by Storn and Price in 1996 (Storn, 1997). It relies on the mutation operation. Mutation and selection operations are used for searching and selecting the most promising regions in the bounded search space. DE uses existing swarm members to construct the trial vector, and a recombination operator to combine better combinations of search parameters for the better solution. Consider an optimization problem to minimize $f(x)$, where $x = (x_1, x_2, x_3, ..., x_D)$ and D specifies the number of dimensions. For a particle population of size N, the population matrix can be written as $x_{n,i}^g = [x_{n,1}^g, x_{n,2}^g, x_{n,3}^g, ..., x_{n,D}^g]$, where g denotes the generation and $x = 1, 2, 3, ...N$.

The algorithm works in four different steps as given below.

Initializing the population: At the beginning, population is generated randomly between lower and upper bounds. Solution vectors are generated randomly at the start.

$x_{n,i} = x_{n,i}^L + rand() * \left(x_{n,i}^U - x_{n,i}^L \right)$, where $i = 1, 2, 3, ...N$, x_i^L is the lower bound and x_i^U is the upper bound of the variable x_i. Subsequent steps will improve the search by applying the mutation operation.

Mutation: A target vector called *mutant vector* and created by applying mutation operation. From each parameter vector, it selects three other vectors $x_{r1,n}^g$, $x_{r2,n}^g$, and $x_{r3,n}^g$ randomly. Then it finds the weighted difference between them and adds the computed difference to the third.

$$v_n^{g+1} = x_{r1,n}^g + F\left(x_{r2,n}^g - x_{r3,n}^g\right) n = 1, 2, ..., N \tag{9}$$

and v_n^{g+1} is called the donor vector. F takes values between 0 and 1.

Crossover: A trial vector v_n^{g+1} is developed from the target vector $x_{rn,i}^g$, and the donor vector.

$$v_n^{g+1} = \begin{cases} u_{n,i}^{g+1} \text{ if } rand() \leq C_p \text{ or } i = I_{rand} \text{ } i = 1, 2, ..., D \text{ and} \\ x_{n,i}^g \text{ if } rand() \leq C_p \text{ and } i \neq I_{rand} \text{ } D = 1, 2, ..., N \end{cases} \tag{10}$$

where I_{rand} is an integer random number between $[0, D]$

Selection: Computes the difference between the vectors $x_{n,i}^g$ and $x_{n,i}^g$ and selects the minimal value function for the next generation.

$$x_n^{g+1} = \begin{cases} u_{n,i}^{g+1} & \text{if } f\left(u_{n,}^{g+1}\right) < \left(x_n^g\right) \\ x_n^g & otherwise \end{cases} \tag{11}$$

The DE algorithm stores the identified best solution in its population by rendering the increase in speed of convergence in PSO.

5.2.4.2 The Concept of CPSO

Composite particle swarm optimization (CPSO) applies the combination learning strategy to increase the convergence speed. It defines two swarm populations. One is particle population for objective function minimization and the other is a DE parameter population for increasing the performance of the algorithm by minimizing the convergence time.

Let $S_t = \{x_1^t, x_2^t, ..., x_i^t, \}$ be the particle population defined on a specific problem in the search space. The CPSO updates the velocity vector and the position vectors of the particles in the swarm as given in the Equations (12) and (13).

$$v_{i,j}^{t+1} = w v_{i,j}^t + c_1 r_{1,j}(t)\left(p_{i,j}(t) - x_{i,j}(t)\right) + c_2 r_{2,j}(t)\left(g_{i,j}(t) - x_{i,j}(t)\right) \tag{12}$$

$$x_i^{t+1} = x_i^t + v_i^{t+1} \tag{13}$$

Where $i = 1, 2, 3, ..., N$ and $j = 1, 2, 3, ...N$. The term w is called *inertia weight*. The term c_1 is the *cognitive parameter* and the term c_2 is the *social parameter.* The terms r_1 and r_2 are the *random vectors.* The terms p_i and g_i are the local and global best locations in the swarm.

For a given swarm S_t, CPSO defines a DE population $P = k_1, k_2, ..., k_m$ in a three-dimensional search space having vector of parameters as given in Equation (14).

$$k_m = w_m, c_{1m}, c_{2m}, \; where \; m = 1, 2, 3, ..., m \tag{14}$$

The swarm S_t is updated and evaluated for the individual parameter k_m in Equation (14). The functional value of the individual parameter k_m is adopted from the functional value of the global best particle in the particle population. Mutation and crossover operations are applied to update the DE population. The procedure continues with the main swarm updates in CSPO until termination criteria have been met by the algorithm.

5.2.5 Guaranteed Convergence PSO (GCPSO)

Particle swarm optimization (PSO) is an effective optimization technique for global search, but it often suffers from premature convergence when applied in solving multimodal optimization problems. The reason is that, in global PSO procedure, the local best position of the particle coincides with the global best position. At this moment, the velocity of the particle depends only on the inertia weight component of the velocity vector. Once the particle approaches its global best solution, the velocity of the particle becomes zero and all the other particles stop moving. This behaviour in global PSO neither has the guarantee of converging at the global optima nor at local optima. It only guarantees about the best solution that has been found until the current iteration. Van Den Bergh (Hernandez et al., 2007) has introduced a new approach called guaranteed convergence PSO (GCPSO).

The standard PSO updates the global velocity vector by considering the inertia weight as given in the Equation (15).

$$v_{i,j}^{t+1} = w v_{i,j}^t + c_1 r_{1,j}(t) \left(p_{i,j}(t) - x_{i,j}(t) \right) + c_2 r_{2,j}(t) \left(g_{i,j}(t) - x_{i,j}(t) \right) \tag{15}$$

The term w is called the *inertia weight.* The term c_1 is the *cognitive parameter,* and the term c_2 is the *social parameter.* The algorithm's convergence properties are determined by these parameters. For the constricted version of PSO, the following restrictions are proposed.

Inertia weight in the equation is used to determine how the previous velocity of the particle influences the velocity of the particle in the next iteration. When $w=0$, the velocity of the particle depends on only p_i and g_i of the particle. Instantly the particle might change its velocity if it moves away from the known location. This makes the lower values of inertia weight suitable for local search in PSO. When $w=1$, the high inertia weight favors the particles in exploration. When $w=0$ and $p_i=g_i$, all the particles stop moving and converge. This may lead to premature convergence in PSO. To address this problem, GCPSO modifies the rule to update the velocity of the best particle of the swarm as given in Equation (16). It modifies the rule to update the velocity of the best particle of the swarm as given in Equation (16).

$$v_{i,j}^{t+1} = w v_{i,j}^t + g_{i,j}(t) - x_{i,j}(t) + \rho^t r_j \tag{16}$$

where r_j denotes a sequence of random numbers varying in the range $[-1, 1]$. ρ^t is a scaling factor determined using Equations (17) and (18).

$$\rho^t = 1.0 \tag{17}$$

$$\rho^{t+1} = \begin{cases} 2\rho^t & if\ successes > s_c \\ 0.5\rho^t & if\ failures > f_c \\ \rho^t & otherwise \end{cases} \tag{18}$$

where s_c and f_c are tunable threshold parameters. When there is an improvement in the personal best position of the best particle, the successes counter is incremented and the failures counter is set to 0. When the best particle gets changed, both successes and the failure counters are set to 0.

This makes GCPSO succeed in making the best particle perform directed random search around its best position in the search space. This leads the swarm toward guaranteed convergence with an optimal solution.

5.2.6 Cooperative PSO

The standard PSO algorithm finds the optimal solution for a problem in an n-dimensional search space by continuously changing the position and velocity of the particles, using their knowledge on *pBest* and *gBest* and also through information sharing among swarms using different topologies. The idea of cooperative PSO originated from the cooperative co-evolutionary genetic algorithm. The standard PSO technique helps in arriving at the optimal solution gradually, but it can exhibit a 'two steps forward, one step back' kind of behaviour (Huisheng & Yanmin, 2011) in some of its components. As a result, some components are attracted toward the optimal solution, and some components head off from the optimal solution. To address this issue, Van Den Bergh and Engelbrecht (2004) developed a cooperative version of the PSO algorithm, called the CPSO algorithm, which is based on decomposition of search space and exchange of information between search spaces to reach an optimal solution. CPSO includes two versions, CPSO-S and CPSO-H.

5.2.6.1 CPSO-S

The CPSO-S approach is called 'the cooperative split PSO approach' (Unger, Ombuki-Berman, & Engelbrecht, 2013). In this approach, the solution vector space is split into smaller vectors or sub-spaces. Each subdivision of space is optimized by an individual swarm. Every solution found by the best particle of each swarm contributes to the overall solution vector. The fitness value for a certain particle i that exists in a swarm j is determined by the solution vector used with that particle and also by the best particles that exist in all the other swarms. The CPSO-S approach is shown in Figure 5.3.

5.2.6.2 CPSO-H

Hybrid CPSO (CPSO-H) is a cooperative approach that is a hybrid of the CPSO-S technique and PSO, introduced in papers (Van Den Bergh & Engelbrecht, 2001, 2004), and it was called CPSO-H. The search occurs in two stages sequentially. Stage 1 employs the CPSO-S technique, and Stage 2 employs the classic PSO algorithm. One stage is run per iteration, and it advances the best solution that it has found to the next stage. The great advantage of this approach is that it exploits the property of fast convergence of CPSO-S in first stage and the PSO's ability to escape pseudo-minimizers in the second stage. This approach is shown in Figure 5.4.

5.2.7 Niching PSO

Niching is a method of cluster formation based on the concept of 'fitness sharing.' It divides the main problem into smaller sub-problems and finds the solutions in parallel. When introduced in a search algorithm, it divides

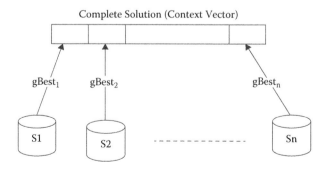

FIGURE 5.3
CPSO-S.

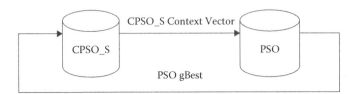

FIGURE 5.4
CPSO-H.

the entire search space into smaller regions, called niches, and applies search on those regions in parallel. Niching is mainly applied in multi-modal optimization problems. It works in sequential or in parallel mode. The sequential method mainly focuses on exploration by which the optimality drives toward a single global solution which is not the main concern of multimodal problem. The parallel niching makes the search algorithm generate multiple niches and identify more refined optimal solutions in the local neighbourhood. The best solutions from each niche will participate in finding the global solution.

In the year 2002, Brits, Engelbrecht, and Van Den Bergh (2002) introduced the nbest PSO, which incorporated parallel niching in a particle swarm. The aim of nbest PSO was to find *n*-parallel solutions for a problem based on the local neighbourhood determined by spatial proximity. Subsequent advents in niching by the same authors led them to derive a more efficient multimodal optimization algorithm called niching PSO. The swarm contains isolated particles by having empty neighbourhood at the beginning. Niching PSO uses the cognition-only model (Eberhart & Kennedy, 1995) to generate the sub-swarms from the main swarm. The absence of the social component in the swarm makes each particle to perform a local search. GCPSO (described in Section 5.2.5) is used to train the sub-swarms, which ensures the convergence to a local point. The procedure for niching PSO is illustrated in the algorithm given in the Figure 5.5.

Niching PSO considers the radius of the sub-swarm as one of the parameters to decide whether or not to absorb the new particle into the sub-swarm and to merge more than one sub-swarm. The steps followed in the algorithms above are elaborated on in the following section.

Main swarm initialization: Particles are distributed randomly in the search space using the *Faure-sequence* (Deb & Goldberg, 1989). The velocity equation is updated using the following equation:

$$v_{i,j}^{t+1} = wv_{i,j}^t + c_1 r_{1,j}(t)\left(p_{i,j}(t) - x_{i,j}(t)\right) + c_2 r_{2,j}(t)\left(g_{i,j}(t) - x_{i,j}(t)\right) \qquad (19)$$

Main Swarm Initialization: Initialize the main particle swarm using *Faure-sequence*

Repeat

          Main swarm training: Train main swarm using Cognition-only Model

              Update the fitness of each particle

      Identify sub swarms:

              For each sub swarm, update neighbourhood radius

                  Apply GCPSO on each sub swarm

                  Update neighbourhood radius of the sub swarm

              End For

      Absorption of the particle into a sub swarm:

              If a particle *p* from the main swarm comes into sub swarm radius,

                  Absorb the particle into the sub swarm

              End if

      Merging of two sub swarms:

              If two sub Swarms overlap in the defined neighbourhood radius,

                  Merge those two sub swarms

              End if

      Create new niche from main swarm:

              For each particle p in the main swarm do

                  If p meets the partitioning criteria then

                      Create a new sub swarm

                  End if

              End For

Until stopping criteria is met

End procedure

FIGURE 5.5
Niching PSO algorithm.

Where x_i denotes the current position and p_i denotes the personal best position of the particle, v_i denotes the current velocity of the particle, and g_i denotes the global best position of the particle.

Main swarm training: The cognition-only model (Eberhart & Kennedy, 1995) is used to train the particle swarm in the global search space. Only the personal best position is used to make every particle do individual search without depending on the social component of the swarm.

$$v_{i,j}^{t+1} = wv_{i,j}^t + c_1 r_{1,j}(t)\left(p_{i,j}(t) - x_{i,j}(t)\right) \tag{20}$$

This arrangement allows each particle to perform only a local search.

Identification of niches: If it has been observed that there is not much difference in the fitness of the particle over iterations; a new niche with that particle has been created. The local neighbourhood for the niche has been created as per Equation (21).

$$R_j = max\left\{\|S_{x_{j,g}} - S_{x_{j,i}}\|\right\} \tag{21}$$

Where g denotes the global best particle in the sub S_j having satisfied with $i \neq j$.

Absorption of the particle into a sub-swarm: A particle xi is absorbed into sub-swarm if it satisfies the condition given in the Equation (22).

$$\|x_i - S_{x_{j,g}}\| \leq R_j \tag{22}$$

Where R_j is the radius of the local neighbourhood.

Merging of two sub-swarms: If any two swarms overlap with each other with respect to the neighbourhood radius, then they are merged into single sub-swarm, i.e., sub-swarm S_{j1} and the sub-swarm S_{j2} is merged if the condition in the Equation (23) is satisfied.

$$\|S_{x_{j1,g}} - S_{x_{j2,g}}\| < \left(R_{j1} - R_{j2}\right) \tag{23}$$

When they are merged, social information about the niche is shared between the swarms, avoiding superfluous traversal of the search space.

5.2.8 Quantum PSO (QPSO)

The Standard PSO has a common problem of premature convergence that leads to optimization stagnation. PSO fails to guarantee global convergence, as analyzed by Van Den Bergh (2002). The quantum particle swarm algorithm (QPSO) was first proposed by Sun (2004); he applied the laws of quantum mechanics to the particle swarm optimization algorithm. We can say that classical PSO follows Newtonian mechanics because the path of the particle is determined by its position vector x_i and velocity vector v_i. In QPSO, we consider that the particles exhibit quantum behaviour in the search space, then x_i and v_i of a particle cannot be determined simultaneously, based on principle of uncertainty. This improves the search ability, convergence speed, and accuracy of the optimal solution when compared to classical PSO.

State of the particle is depicted by a wave function $\varphi(x, t)$ in QPSO instead of the position and velocity used in classical PSO. The probability density function $|\varphi(x, t)^2|$ gives the probability of the particle in position x_i and this function depends on the potential field the particle lies in. The particles move based on Equation 24.

$$x(t+1) = \begin{cases} p + \beta * |m_{best} - x(t)| * I_n(1/u) & if \ k \geq 0.5 \\ p - \beta * |m_{best} - x(t)| * I_n(1/u) & if \ k \leq 0.5 \end{cases} \quad (24)$$

where

$$P = (c_1 P_{id} + c_{12} P_{gd})/(c_1 + c_2) \quad (25)$$

$$m_{best} = \frac{1}{M}\sum_{i=1}^{M} P_i = \left[\frac{1}{M}\sum_{i=1}^{M} P_{i1}, \frac{1}{M}\sum_{i=1}^{M} P_{i2}, ..., \frac{1}{M}\sum_{i=1}^{M} P_{id}\right] \quad (26)$$

m_{best} is the mean best of the population. It is the mean of the best positions of all particles. u, k, c_1 and c_1 are uniformly distributed random numbers in the interval [0, 1]. The parameter β is called *contraction-expansion* coefficient. The overall steps followed in QPSO are given in Figure 5.6.

5.2.9 Constrained PSO (COPSO)

Constrained PSO (COPSO) is a relatively new approach introduced for finding the solution for constrained optimization problems. It has introduced a new approach to prevent premature convergence in PSO by adding a constraint handling mechanism. Hu and Eberhart (2002) have

---

Initialize the swarm

Do

       Calculate m_{best} by equation (26)

       Update particles position using equation (24)

       Update *pBest*

       Update *gBest*

While iteration reaches maximum count

FIGURE 5.6
Pseudo code for QPSO.

proposed a feasibility-preserving strategy to determine the best particle in the search space in a constrained environment. The experimental analysis done by G. Coath (Coath & Halgamuge, 2003) has shown that there is a need for some diversity control in the swarm. Based on the research updates, Arturo Hernandez et al. (2007) have introduced a new scheme for constrained PS. It adds two new perturbation operations and a selfless singly linked local neighbourhood topology to identify more feasible solutions in the global search space by encouraging continued inter-swarm communication in the constrained environment. The algorithm defines three different components in three different stages of the optimization process.

Stage 1: Define a selfless model for local neighbourhood topology
Every particle in the population is initialized with a permanent label, which is independent of its geographical location and is used to define the neighbourhood of the particle in search space.

Each particle i, is made to have $n/2$ singly linked particles in both forward and reverse directions, forming a local neighbourhood. The best particle is identified in the local neighbourhood by inspecting the particles $i+1, i+2, i+3, ..., i+n/2$ and $i-1, i-2, i-3, ..., i-n/2$ defined by a singly linked ring topology (selfless model) given in the Figure 5.7. The particle i is not considered a member in the neighbourhood and hence, this is called a selfless model.

The neighbourhood algorithm for the selfless model is given in Figure 5.8.

For improving the local best algorithm (exploitation), COPSO uses an external perturbation procedures, which increases the diversity in the

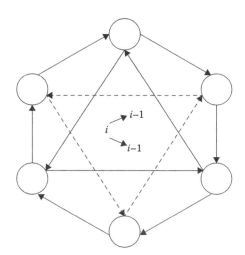

FIGURE 5.7
Singly linked ring topology.

1. Set *step* = 1

2. Set *switch* = 1 (Pick the neighbours either in clockwise or in anticlockwise

direction)

3. Identify new neighbour particle *i* + *switch* * *step*

4. Increment *step* = *step* + 1

5. Compute *switch* = −*switch*

FIGURE 5.8
Neighbourhood algorithm for the selfless model.

swarm and guides the swarm toward the best spots without altering the self-organization capacity of the particle. Perturbation operators will change only the memory of the local best (P_i) particle by allowing the particles to fly around the search space by maintaining diversity.

Stage 2: Perturbation

- **C-Perturbation**

In the second stage, COPSO adds C-perturbation, which is applied on the members of the local best population to get a group of temporal particles named *Temp*. Temp members are compared with their parent particle and the local best, and the winner is selected for the next iteration.

- **M-Perturbation**

In the third stage, next level of perturbation is introduced by adding a small random number to the design variables, which makes the variable deviate from the current direction and leads toward exploration of a more promising region. Every member of the local best is then compared with its parent and the particle best, and the winner is selected for the next iteration.

Stage 3: Constraint Handling

Constraint handling in COPSO was introduced by K. Deb (Deb & Goldberg, 1989) based on a tournament selection model. Any particle belonging to a constrained search space has to first comply with the set of constraints and is then allowed to comply with the function value. COPSO applies this kind of tournament selection method to identify the local best particle and the personal best particle among the given set of feasible particles. Tournament selection proceeds through three different stages.

Stage 1: The particle with better function value is selected.

Stage 2: In the case where there is one feasible and other unfeasible particles, the feasible one is selected.

Stage 3: If both the particles are unfeasible, then the one that has the lowest sum of constraint violations is selected.

5.3 Particle Swarm Optimization (PSO) – Advanced Concepts

5.3.1 PSO Algorithm Adaptation to the Bird-Flocking Scenario

PSO algorithm simulates the behaviour of bird flocking. A flock of birds randomly fly over a region in search of food. Assume there is only one piece of food in the region of search. Initially no bird is aware of the exact location of food. The bird that is nearest to the food alerts the other birds with a loud chirp, and they start flying in his direction. By chance if the location of food is closer to any other birds than that of the first one, it alerts the others with a louder chirp, and the others change the direction and start flying toward him. Over a number of iterations, they come to know how distantly the food is located and get even closer to the food. Following the bird that is nearest to the food will be the best strategy to reach the food. This pattern continues until the food is found by any one of the birds. In PSO, the candidate solutions represent the birds in the search space. The candidate solutions are referred as particles. Each of the particles possesses some fitness value related to it that corresponds with the loudness in the chirp of the birds. The fitness value is calculated by the fitness function of the problem to be optimized, and the particles have velocities that control the flying of the particles to reach the solution. The particles move through the problem space by chasing the current optimal solution, i.e., the first bird that chirps louder and is closest to the food. Particles refresh themselves with their velocity, and they also possess memory to keep track of the history of their current location and also the best locations with respect to other members of the swarm. The next iteration takes place only after all particles have been relocated.

After a number of iterations, the particle adjusts the values of a group of variables that is associated with it, and finally at some point the whole swarm gets nearer to the particle whose value is nearest to the target at any given time.

5.3.2 PSO Algorithm

The PSO algorithm consists of three steps, namely, generating particles' position and velocities, velocity update, and position update. PSO generates and distributes the particles randomly in a N dimensional search space as in the equation Equation (27).

$$x_{min} + rand\,(x_{max} - x_{min})$$ (27)

Where, x_{min} denotes the lower limit of the search space and x_{max} denotes the upper limit of the search space.

Each individual in the particle swarm is initialized with particle position P_i and particle velocity v_i. The best position of the particle in the local neighbourhood is initialized as *pBest* and the best position among all the particles among the entire population is initialized as *gBest*. Each particle in the swarm is updated with its position P_i and the velocity v_i as given in Equation (28).

$$p_i^{t+1} = p_i^t + v_i^{t+1}$$ (28)

Where P_i denotes the particle position, v_i denotes the particle velocity and t denotes the iteration in the algorithm.

The new position of the particle P_i is compared with the local best position of the particle (*pBest*). If the new position of the particle P_i found to be better than the *pBest* position, new position P_i is assigned to the local best position *pBest*. The computed *pBest* value is compared with the global best position of the particle *gBest*. If *pBest* found to be better than the *gBest*, then *gBest* is assigned with the new *pBest* value. The procedure continues until the termination of the criteria are met.

Algorithm adds two random variables $r1, r2$ to maintain uniform distribution of the particles in local and global search space and avoids the algorithm from entrapment in local optima. Algorithm also adds three weight factors namely *inertia weight* denoted by w, particle's *self-confidence factor*, $c1$ and the swarm confidence factor $c2$, which has an influence on search direction, the particle's motion, and the swarm influence. PSO updates its velocity vector using the formula given in the Equation (29).

$$v_{i,j}^{t+1} = w v_{i,j}^t + c_1 r_{1,j}\big(p_{i,j}(t) - x_{i,j}(t)\big) + c_2 r_{2,j}\big(g_{i,j}(t) - x_{i,j}(t)\big)$$ (29)

The pseudo-code for the algorithm is given in the Figure 5.9.

5.3.3 Global Best PSO

The global best PSO reflects a fully connected network topology (Section 5.4.5) so that particles get social information from all particles in the entire swarm (Engelbrecht, 2007; Li & Deb, 2010). The global best PSO (or *gBest* PSO) is a method in which each particle's position is determined and updated by the

```
For each particle
            Initialize particle position in a N dimensional search space
Do
For each particle
            Calculate particle fitness value
            If the fitness value is better than pBest
                        Set pBest = current fitness value
            If pBest is better than gBest
                        Set gBest = pBest
For each particle
            Calculate particle velocity
            Use gBest and Velocity to update particle data
Until maximum iterations or minimum error criteria meets
```

FIGURE 5.9
Pseudo-code for PSO.

best-fit particle in the entire swarm (*gBest*). In *gBest* PSO, each particle considers the entire particle population as its neighbourhood. It searches for the best particle by exploring the global search space. Particles get diversified by the influence of the social component ($c2$)included in the algorithm.

5.3.4 Local Best PSO

The local best PSO (*pBest* PSO) method only allows each particle to be influenced by the best-fit particle selected from its adjacent particles. This reflects a ring topology (Section 5.4.8). Since the particle exchanges information only with its adjacent neighbours, it gains only a local knowledge of the swarm and hence is called local Best PSO (Engelbrecht, 2007; Li & Deb, 2010).

5.3.5 Neighbourhood Topologies

The neighbourhood of each particle influences every particle's movement, and it acts as a very important decision factor that influences the extent of interaction among particles in the swarm. A compact neighbourhood may help in achieving good quality of solutions, but it has a drawback of slower convergence due to less interaction (Talukder, 2011). For a vast neighbourhood, the convergence will be faster, but it may lead to premature convergence. This problem can be addressed by starting the search process with small neighbourhoods and scaling it higher over time. Some neighbourhood topologies are analyzed below:

Figure 5.10 illustrates the fully connected network topology. This topology is called *as gBest* PSO topology resembles a social network, where every particle is connected with one another and can communicate with every other particle in the network. The advantage of this topology is faster convergence, but there are chances of getting stuck in local minima.

Figure 5.11 shows the ring topology. This topology is called as *lBest* PSO topology, where each particle has access only to its adjacent neighbours. When a particle finds a better solution, it advances it to its adjacent neighbours, and these two adjacent neighbours push the solution to their adjacent ones, until the last particle receives the better solution. The advantage of this topology is it covers most of the search space, but the convergence is slower.

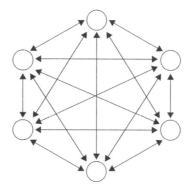

FIGURE 5.10
Fully connected network topology.

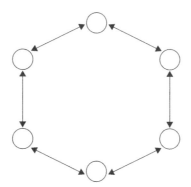

FIGURE 5.11
Ring topology.

Figure 5.12 illustrates the wheel topology in which the particles have no link with other particles except the focal particle, and all information exchange occurs only through the focal particle. The focal particle calculates the new position to reach the best particle by making a comparison of every particle's best performance in the swarm and passes the information to other particles.

Figure 5.13 shows a four-clusters topology in which four clusters are associated with one edge linking opposite clusters and with two edges linking neighbouring clusters. If a cluster finds a best solution, it communicates the

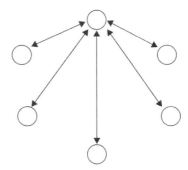

FIGURE 5.12
Wheel topology.

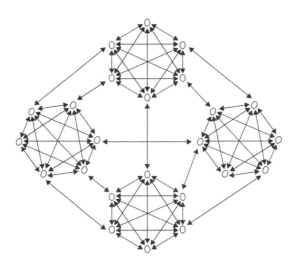

FIGURE 5.13
Four-clusters topology.

information to the other connected clusters through only one particle in the cluster, called the informer.

5.4 Applications of PSO in Various Engineering Domains

5.4.1 Signal Processing

Application of PSO has a great deal in the field of signal processing. Some of them include flatness signal pattern recognition, network signal processing and intrusion detection, adaptive filtering, weak signal detection, time frequency analysis and non-stationary power signal classification, blind detection in the frequency domain, analogue filter tuning for antenna matching, speech coding, de-noising of ECG signal, and the fault diagnosis method of radar signal processing system.

5.4.2 Neural Networks

There are various PSO applications in the field of neural networks. The applications include the areas like design of artificial neural networks (ANN), fuzzy neural network controller, feed forward neural network training, cellular neural networks, neuron controllers, weight optimization for a neural network, and the PSO optimized Hopfield neural network.

5.4.3 Robotics

A substantial proportion of the applications of PSO exist in the field of robotics. They include voice control of robots, path planning, swarm robotics, obstacle avoidance, unmanned vehicle navigation, soccer playing, robotic map building, Multi-robot learning, and parameter estimation of Robot dynamics.

5.4.4 Power System and Plants

A broad range of PSO applications deals with power systems. Specific applications include power loss minimization, power system restoration, operations planning, power system security, optimal power pooling for a multiple-area power system, secondary voltage control, STATCOM power system, optimal load dispatch, and optimal power flow.

5.4.5 Electronics and Electromagnetics

Here, we list the applications of PSO in the field of electronics and electromagnetics. These include conductors, high-speed CMOS IC design and optimization, voltage flicker measurement, electromagnetic shape design, power electronics circuit design, solar power tracking, support vector machine, leakage power reduction in VLSI, and multimodal electromagnetic problems.

5.4.6 Combinatorial Optimization

PSO helps in dealing with optimization of combinatorial problems. These include applications such as the blocking flow shop scheduling problem, urban planning, the travelling salesman problem, minimum spanning trees, packing and knapsack, vehicle routing problem, the *n*-queens problem, and layout optimization.

5.4.7 Electricity Networks and Load Distribution

PSO helps in dealing with the problems of electricity networks and load distribution. Some of the applications include micro grids, minimizing power loss, economic dispatch problem, smart load management, transmission network planning, distribution network reconfiguration and expansion, congestion management, and voltage profile improvement of the distribution system.

5.4.8 Scheduling

PSO helps in optimal scheduling applications. Some of the applications include anesthesiology nurse scheduling, scheduling nonlinear multimedia services, and integration of process planning and scheduling, flow shop scheduling, power generation scheduling, grid scheduling, guidance path scheduling, hydrothermal scheduling, staff scheduling, timetable scheduling, sleep scheduling for wireless sensor networks, and cloudlet scheduling.

5.4.9 Analysis of Image and Video

Some of the image analysis applications are medical image segmentation, watershed image analysis, image denoising, colour image enhancement, image edge detection, optimal feature selection, medical image analysis, image hiding scheme, pedestrian detection and tracking, identification of cancer in CT images, and so on. A few video applications are concept detection of video data, video summarization, tracking of an object in video stream, finding the best matching block, body posture tracking, fractal mobile video compression, and video super-resolution.

5.4.10 Prediction and Forecasting

The major areas where PSO is applied are meteorological data study and forecasting, stock price forecasting, crystal structure prediction, time-series forecasting of atmospheric pollutant concentration levels, short and mid-term wind power forecasting, electricity demand forecasting, rainfall fore-casting, predictions of air pollution, forecasting natural gas consumption, sales growth rate forecasting, and so on.

5.5 Conclusion

In this chapter, we have made efforts to give an idea of the origin of PSO and the basic PSO algorithm. Basic particle swarm optimization has both merits and disadvantages. To overcome the inadequacy of PSO in addressing various problems, several basic variants of PSO are also brought into the limelight. PSO is very attractive for the researchers because the algorithm clearly shimmers for its simplicity. It is considered for its compliance with different application sectors and association quality with other approaches. The PSO gave good results in the areas where it has been applied, and the applications are exponentially growing, among which some areas were brought into concern.

References

Brits, R., Engelbrecht, A. P., & Van Den Bergh, F. (2002). A niching particle swarm optimizer. In *Proceedings of the 4th Asia-Pacific Conference on Simulated Evolution and Learning (SEAL'02)* (L. Wang, K. C. Tan, T. Furuhashi, J. H. Kim, and X.Yao, eds), vol. 2, Singapore, November, pp. 692–696.

Coath, G., & Halgamuge, S. K. (2003). A comparison of constraint-handling methods for the application of particle swarm optimization to constrained nonlinear optimization problems. In *Proceedings of the 2003 Congress on Evolutionary Computation*, December, pp. 2419–2425. IEEE.

Coello Coello, C. A., & Sierra, M. R. (2004). A study of the parallelization of a coevolutionary multi-objective evolutionary algorithm. In C. A. Coello Coello (Ed.), Margarita Reyes Sierra Mexican International Conference on Artificial Intelligence, MICAI 2004: MICAI 2004: Advances in artificial intelligence (pp. 688–697).

Dawkins, R. (1976). *The Selfish Gene*. Oxford: Clarendon Press.

Deb, K., & Goldberg, D. E. (1989). An investigation of niche and species formation in genetic function optimization. In *Proceedings of the Third International Conference on Genetic Algorithms* (pp. 42–50). San Francisco, CA: Morgan Kaufmann Publishers Inc.

Eberhart, R. C., & Kennedy, J. (1995). A new optimizer using particle swarm theory. In: *Proceedings of the Sixth International Symposium on Micromachine and Human Science*. IEEE Service Center, 39–43.

Eberhart, R. C., & Shi, Y. (1998). Comparison between genetic algorithms and particle swarm optimization. In *International Conference on Evolutionary Programming EP: Evolutionary Programming VII*, pp. 611–616.

Engelbrecht, A. P. (2007). *Computational Intelligence: An Introduction* (Second edition ed.). John Wiley.

Heppner, F., & Grenander, U. (1990). A stochastic nonlinear model for coordinated bird flocks. In Krasner S (Ed.), *The Ubiquity of Chaos*. Washington, DC: AAAS Publications.

Hernandez, A. A., Muñoz Zavala, A. E., Villa Diharce, E., & Botello Rionda, S. (2007). COPSO: Constrained optimization via PSO algorithm (*Comunicación Técnica* No I-07-04/22-02-2007 (CC/CIMAT)).

Hu, X., & Eberhart, R. C. (2002). Solving constrained nonlinear optimization problems with particle swarm optimization. In *Proceedings of the World Multi-conference on Systemics, Cybernetics and Informatics*, Orlando, FL, pp. 122–127.

Huisheng, S., & Yanmin, Z. (2011). An improved cooperative PSO algorithm. In *2011 International Conference on Mechatronic Science, Electric Engineering and Computer*, August 19–22, Jilin, China.

Jones, K. O. (2005). Comparison of genetic algorithm and particle swarm optimization. In *International Conference on Computer Systems and Technologies - CompSysTech*.

Kennedy, J., & Eberhart, R. (1995). Particle swarm optimization. In *Proceedings of the IEEE International Conference on Neural Networks*, Vol. 4, pp. 1942–1948, December.

Li, X., & Deb, K. (2010). Comparing lbest PSO niching algorithms using different position update rules. In *Proceedings of the IEEE World Congress on Computational Intelligence*, Spain, pp. 1564–1571.

Millonas, M. M. (1993). Swarms, phase transitions, and collective intelligence. In LangtonCG (Ed.) *Proceedings of ALIFE III. USA*: Santa Fe Institute, Addison-Wesley.

Montes De Oca, M. A., Stutzle, T., Birattari, M., & Dorigo, M. (2007). Frankenstein's PSO: Complete data. http://iridia.ulb.ac.be/supp/IridiaSupp2007-002/.

Moscato, P. (1989). *On Evolution, Search, Optimization, Genetic Algorithms and Martial Arts Towards Memetic Algorithms*. Caltech concurrent computation program 158-79. Pasadena: California Institute of Technology.

Nowak, A., Szamrej, J., & Latané, B. (1990). From private attitude to public opinion: A dynamic theory of social impact. *Psychological Review, 97*, 362. doi:10.1037/0033-295X.97.3.362

Parsopoulos, K. E, Tasoulis, D. E., & Vrahatis, M. N. (2004). Multi-objective optimization using parallel vector evaluated particle swarm optimization. In *Proceedings of International Conference on Artificial Intelligence and Applications (IASTED)*, Innsbruck, Austria.

Parsopoulos, K. E., & Vrahatis, M. N. (2002). Recent approaches to global optimization problems through particle swarm optimization. *Natural Computing, 1*, 235–306. © 2002 Kluwer Academic Publishers. Printed in the Netherlands.

Parsopoulos, K. E., & Vrahatis, M. N. (2004a). On the computation of all global minimizers through particle swarm optimization. *IEEE Transaction on Evolutionary Computation, 8*(3), 211–224.

Parsopoulos, K. E., & Vrahatis, M. N. (2004b). UPSO: A unified particle swarm optimization scheme. In T., Simos & G. Maroulis (Eds.), *Lecture Series on Computer and Computational Sciences* (Vol. 1, 868–873). Zeist, The Netherlands: VSP International Science Publishers.

Petalas, Y. G., Parsopoulos, K. E., & Vrahatis, M. N. (2007). Memetic particle swarmoptimization. *Annals of Operations Research*, *156*, 99–127.

Price, K., Storn, R., & Lampinen, J. (2005). *Differentialvolution: A Practical Approach to Global Optimization*. Berlin, Germany: Springer Verlag.

Reeves, W. T. (1983). Particle systems—a technique for modeling a class of fuzzy objects. *ACM Transactions on Graphics*, *2*(2), 91–108.

Reynolds, C. W. (1987). Flocks, herds, and schools: a distributed behavioral model. *Computer Graphics and Interactive Techniques*, *21*(4), 25–34.

Schaffer, J. D. (1984). *Multiobjective Optimization with Vector Evaluated Genetic Algorithms*. PhD Thesis, Vanderbilt University, Nashville, USA.

Storn, R. (1997). Differential evolution, a simple and efficient heuristic strategy for global optimization over continuous spaces. *Journal of Global Optimization*, *11*, 341–359.

Sun, J. S. (2004). Particle swarm optimization with particles having quantum behavior. *Proceedings of Congress on Evolutionary Computing*, *1*, 325–331.

Talukder, S. (2011). *Mathematical Modelling and Applications of Particle Swarm Optimization by Satyobroto*.

Unger, N. J., Ombuki-Berman, B. M., & Engelbrecht, A. P. (2013). Cooperative particle swarm optimization in dynamic environments. *IEEE Symposium on Swarm Intelligence (SIS)*.

Van Den Bergh, F., & Engelbrecht, A. P. (2001). Effect of swarm size on cooperative particle swarm optimizers. In *Proceedings of the 3rd Annual Conference on Genetic and Evolutionary Computation*.

Van Den Bergh, F., & Engelbrecht, A. P. (2004). A cooperative approach to particle swarm optimization. *IEEE Transactions on Evolutionary Computation*, *8*, 225–239.

Van Den Bergh, F., & Engelbrecht, A. P. (2006). A study of particle swarm optimization particle trajectories. *Information Sciences*, *176*, 937–971.

6

Artificial Bee Colony, Firefly Swarm Optimization, and Bat Algorithms

Sandeep Kumar and Rajani Kumari

CONTENTS

6.1 Introduction

The ABC simulates foraging behaviour of real honey bees to solve multi-model and multi-dimensional problems. The intelligent behaviour of honey bees attracted various researchers and led them to develop some new strategies. There are a number of algorithms based on bees' behaviour while searching for food, locating partners for mating, social organization, etc. In 2005, Karaboga (2005) first identified that bees follow a systematic process while collecting nectar in order to produce honey. It is stochastic in nature as some arbitrary element plays a crucial role in the process of optimization and also depicts swarming behaviour. Honey bees show extraordinary behaviour while probing for proper food sources and then exploiting nectar from best food sources. Based on behaviour of honey bees, the ABC algorithm is divided into three phases. The ABC algorithm is very efficient in comparison to other competitive algorithms, but it has a few drawbacks, also similarly to other algorithms, such as premature convergence and stagnation. That is the reason that, since its inceptions, it has been modified by a number of researchers. The ABC algorithm is very simple in implementation because it has only three control parameters and follows three simple steps while moving toward optimum.

The firefly algorithm (Yang, 2009) depicts the flashing behaviour of fireflies while searching for a mating partner or prey and to release a warning to enemies. Nature has a large number of firefly species; most live in a tropical environment. Fireflies produce a special type of cold light by a chemical reaction known as Bioluminescence. Bioluminescence happens through the oxidation of an organic substance in living organisms. Apart from the firefly, nature has various living beings that emit light, but the firefly shows a special kind of pattern while flashing. The female fireflies in the same species respond to flashing pattern of male fireflies for mating. In some cases, female fireflies attract male fireflies from other species for hunting. Xin-She Yang developed an intelligent population-based algorithm after observing the flashing pattern and mating behaviour of fireflies in 2008 (Yang, 2009). In (Yang, 2009), it is presumed that all fireflies are unisex and they are fascinated by each other irrespective of their sex. The attractiveness of a firefly depends on brightness of other fireflies. A brighter firefly attracts a less bright firefly, i.e., attractiveness is directly proportional to intensity of luminescence. The beauty of the firefly algorithm is that it gives a better convergence rate in spite of bad initial solutions.

The bat algorithm is based on the echolocation behaviour of bats, especially micro-bats with a fluctuating pulse rate of emission and loudness. The bats produce some sounds (ultrasonic waves) that are reflected by entities in their path and send echoes back to the bats. With the help of these echoes, the bat decides the size of object, the distance from it, and its current state (moving or stationary). The bat traps their respective prey using this information. Xin-She Yang (Yang, 2010a), in 2010, proposed an algorithm on this extraordinary behaviour of micro-bats that they exhibit while searching for prey in dark places. The bat algorithm is very simple in implementation and has few parameters. It gives comparatively fast convergence for highly complex and non-linear optimization problems. Because it switches from exploration to exploitation in the very early stages, though, it sometimes leads to stagnation also. It is highly adaptive and performs auto-skyrocketing in the search space due to the pulse emanation rate and loudness. In the bat algorithm, it is assumed that a virtual bat flies arbitrarily at location (solution) x_i with a velocity v_i and a changeable frequency or wavelength and loudness A_i. When it finds prey, it changes frequency, loudness, and pulse emanation rate r. The bats' search process is strengthened by a local random walk. It continuously searches for best until the termination criteria are met. Basically, the dynamic behaviour of a bat swarm is controlled by a frequency-tuning procedure, and the balance between exploration and exploitation can be controlled by fine tuning parameters.

6.2 The Artificial Bee Colony Algorithm

6.2.1 Introduction

The ABC algorithm is developed by Dervis Karaboga in 2005 (Karaboga, 2005). It mimics the intelligent foraging behaviour of real honey bees. The ABC algorithm was inspired by the extraordinary conduct of honey bees while searching for better-quality food sources. Analogous to natural honey bees, the ABC algorithm divides all bees in three different groups according to their behaviour and nature of task performed. The whole population is composed of three types of bees: employed, onlooker, and scout bee. The employed bees are responsible for searching for new food sources and providing information about food sources to bees that reside in the hive (onlooker bees). Based on information received from employed bees, onlooker bees start exploiting these food sources. If food sources are exhausted due to exploitation, they are considered abandoned and onlooker bees are replaced by scout bees. The bees are continuously trying to improve solutions using a greedy search strategy until the termination criteria are met, and they memorize the best solution established so far. The success of ABC algorithm depends on balance between these two processes. Initialization of swarm also play important role in deciding direction of the solution search process.

The behaviour of natural honey bees can be described by the following parameters:

- Food source
- Bees
- Dance

The bees identify a specific food source (flower) while searching for food. They then start collecting information about that particular food source (i.e., quantity of nectar that may be exploitable from it), its direction, and its distance from hive. A food source is analogous to a solution in a search space. In order to get a better outcome, it is essentially required to use a proper initialization strategy and an exploitation mechanism that can accelerate finding good solutions in the next iteration. The group of bees that voluntarily starts searching for food sources is known as employed bees. These bees collect information about food sources and share it with other bees that are waiting in beehive. This information exchange takes place using a special kind of dance, like the round, waggle, and tremble dance. Bees that observe the dance of employed bees are known as onlooker bees. Employed bees identify the best food sources, which are rich in nectar. Now onlooker bees start exploiting these food sources.

6.2.2 Phases of the Artificial Bee Colony Algorithm

The artificial bee colony algorithm has three phases namely: the employed bee phase, the onlooker bee phase, and the scout bee phase. These three steps are iterated to find optimal solutions after initialization. The key steps of ABC algorithm are revealed in Figure 6.1.

Initialization: The first step in ABC is initialization of parameters (colony size, limit for scout bees, and maximum number of cycles) and setup of an initial population randomly using the following equation.

$$p_{ij} = LB_j + rand(0,1) \times (UB_j - LB_j) \tag{6.1}$$

Where $i = 1, 2..., (Colony\ Size/2)$ and $j = 1, 2, .., D$. D represent dimension of problem. p_{ij} denotes the position of ith food source in jth dimension. LB_j and UB_j denotes the lower and upper boundary values of the search space, respectively. $rand\,(0,1)$ is a randomly selected number in the range $(0,1)$.

Employed Bee Phase: This phase tries to detect superior quality solutions in the proximity of current solutions. If the quality of the fresh solution is better than the present solution, then the position is updated;

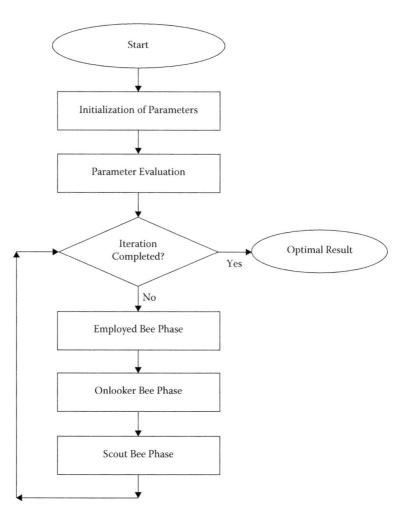

FIGURE 6.1
Flow chart for ABC algorithm.

otherwise, it remains the same. The position of employed bee is updated using the following equation.

$$V_{ij} = p_{ij} + \varphi_{ij} \times \overbrace{(p_{ij} - p_{kj})}^{s}$$ (6.2)

Where φ_{ij} is an arbitrary number in the range [-1, 1], $k = 1, 2..., $ (*Colony Size*/2) is a randomly selected index such that $k \neq i$. In this equation, s denotes step size of the position update equation. A larger step size leads to the skipping of the actual solution, and tiny step size leads to slow convergence.

Onlooker Bee Phase: The onlooker bee selects a food source based on probability of selection and becomes an employed bee. The new employed bee starts searching for an innovative solution in the proximity of the present solution. The probability is computed using the fitness of solution with the help of the following equation:

$$Prob_i = \frac{fitness_i}{\sum_{i=1}^{colony\ Sizze/2} fitness_i}$$ (6.3)

The fitness of a solution must be a function of its value. The fitness of an individual is calculated using the following equation.

$$fitnes\ s_i = \begin{cases} 1/(1 + f(x_i)) & f(x_i) \geq 0, \\ 1 + abs(f(x_i)) & else \end{cases}$$ (6.4)

Scout Bee Phase: An employed bee becomes scout bee when the solution value is not updated, until the predefined threshold limit. This scout bee engenders a new solution instead of the rejected solution using Equation (6.1).

The computational complexity of ABC depends on population size and the dimension of problem. If N is the size of considered population, then each loop (in ABC it has two loops, one for each onlooker and employed bee phase) iterates $N/2$ times (as number of employed bees and onlooker bees are $N/2$), and every phase contains loop over D (dimension of problem). The algorithm stops after meeting the termination criteria (maximum number of iterations). Thus, the run-time complexity of ABC may be denoted by $O(2 \times (N/2) \times D \times Max\_Iteration)$.

6.2.3 Mathematical Proof of the Artificial Bee Colony Algorithm

The ABC algorithm consists of three phases that are iterated after initialization of the population. Consider a sphere function to prove mathematical interpretation of ABC algorithm. The sphere function is continuous, convex, and uni-modal function with d local minima and one global minima.

$$f(x) = \sum_{i=1}^{d} x_i^2, \ x_i \in [-5.12, 5.12] \ \forall i = 1, 2, \ldots, d. \ \text{ and } \ f(x^*) = 0 \ at \ x^* = (0, \ldots, 0)$$

Where d represents the dimension of problem.

Assume that

Dimension (d) = 2,

Size of population (NP) = 6,

Limit (L) = ($NP*D$)/2 = (6*2)/2 = 6

For $d=2$, the function will be $f(x) = x_1^2 + x_2^2$

As per Karaboga (2005), the number of employed bees and onlooker bees should be half of population/colony size. Thus, initialize the locations of three food sources ($NP/2$) of employed bees, arbitrarily using an even distribution in the range (-5.12, 5.12). The randomly initialized solutions are:

$$x = \begin{pmatrix} 2.3212 & 1.1221 \\ -0.3210 & 2.1234 \\ 0.1891 & -0.1466 \end{pmatrix}$$

With the help of these solutions, the value of the functions is computed as shown below:

$$f(x) \text{ values are: } \begin{pmatrix} 6.6471 \\ 4.6118 \\ 0.0573 \end{pmatrix}$$

The fitness function $fitness_i = \begin{cases} 1/(1 + f(x_i)) & f(x_i) \geq 0, \\ 1 + abs(f(x_i)) & else \end{cases}$ used to compute fitness of each solution.

$$\text{Initial fitness vector is: } \begin{pmatrix} 0.1308 \\ 0.1782 \\ 0.9458 \end{pmatrix}$$

The highest fitness (0.9458) in the current swarm for third solution represents the best-quality food source.

Now, apply the employed bee phase and update the solution using greedy selection, if a solution did not get updated, until the specified limit is reached, then use a scout bee for it.

Iteration =1

Employed bee phase

Update the first employed bee (i.e.,$i=1$) using $V_{ij} = p_{ij} + \varphi_{ij} \times (p_{ij} - p_{kj})$.

Assume that $j=0$, $k=1$, and $\varphi_{ij}=0.1234$ are randomly selected.

$$V_{00} = p_{00} + \varphi_{00} \times (p_{00} - p_{10}) = 2.3212 + (0.1234) * (2.3212 - (-0.3210)) = 2.6473$$

$$V_0 = (2.6473, 1.1221)$$

Now compute $f(V_0)$ and the fitness of V_0.
$f(V_0) = 8.2670$ and Fitness$(V_0) = 0.1070$
Now apply greedy selection between the initial solution (x_0) and the new solution (v_0).

$$0.1070 < 0.1308$$

It can be observed that the fitness of the new solution is not better than that of the old solution; thus, it couldn't be updated. In this case, update the trial counter (counter for scout) because the solution remains unchanged.
Repeat this process for all employed bees (in this case, employed bee are three).
After first iteration

$$x = \begin{pmatrix} 2.3212 & 1.1221 \\ -0.3210 & 2.1234 \\ 0.1870 & -0.1466 \end{pmatrix}$$

$f(x)$ values are: $\begin{pmatrix} 6.6471 \\ 4.6118 \\ 0.0565 \end{pmatrix}$

Fitness vector: $\begin{pmatrix} 0.1308 \\ 0.1782 \\ 0.9465 \end{pmatrix}$

Now compute probability of each solution using their fitness by

$$Prob_i = \frac{fitness_i}{\sum_{i=1}^{colonySizze/2} fitness_i}$$

Probability vector: $\begin{pmatrix} 0.1042 \\ 0.1419 \\ 0.7539 \end{pmatrix}$

Onlooker Bee Phase
Generate new solutions for onlooker bees using the existing solution selected based on probability. Update the first onlooker bee using the position update equation, similar to the employed bee. Randomly select indices j and k.
Assume that $i = 2$, $j = 0$, and $k = 1$. Then

$$V_{20} = p_{20} + \varphi_{20} \times (p_{20} - p_{10}) = 0.1870 + (0.5633)(0.1870 - (-0.3210)) = 0.4632$$

The new solution $V_2 = (0.4632, -0.1466)$

Compute the function value and fitness for v_2.

$f(V_2) = 0.2454$ and Fitness $(V_2) = 0.8030$

Apply the greedy selection among old and new solutions, 0.8030 <0.9465; the solution couldn't be improved, increase the trial counter.

Similarly, apply this phase for all onlooker bees. After the onlooker bee phase

$$x = \begin{pmatrix} 2.3212 & 1.1221 \\ -0.3210 & 2.1234 \\ 0.1856 & -0.1466 \end{pmatrix}$$

$f(x)$ values are: $\begin{pmatrix} 6.6471 \\ 4.6118 \\ 0.0559 \end{pmatrix}$

Fitness vector: $\begin{pmatrix} 0.1308 \\ 0.1782 \\ 0.9470 \end{pmatrix}$

After the onlooker bee phase, remember the best solution found so far = $(0.1856, -0.1466)$

Scout Bee Phase

Trail counter status is

$$\begin{pmatrix} 1 \\ 1 \\ 0 \end{pmatrix}$$

Until now, there was no abandoned solution because the limit for scout is 6. An abandoned solution is replaced by an arbitrary solution if the trial counter is higher than 6.

This complete process repeated until the termination criteria are accomplished.

6.2.4 Recent Modifications in the Artificial Bee Colony Algorithm

The ABC is one of the most interesting algorithms for researchers who are working in the field of optimization. Modifications in the ABC are based on the introduction of new local search strategies, hybridization of ABC with other population base strategies, fine tuning of control parameters, and the introduction of new control parameters. Most of the modifications concentrate on the position update process and local search process. The ABC has undergone various modifications since its inception, and a few of them are discussed here.

Hancer, Xue, Zhang, Karaboga, and Akay (2018) anticipated a modified version of the ABC for multi-objective optimization problems. Here, the ABC was combined with genetic operators and a non-dominated sorting system. Binary and continuous, two different variants of the multi-objective ABC were developed and implemented. The proposed approaches were used for a multi-objective feature selection tactic for classification, and results show that binary multi-objective ABC is far better than classical approaches. Xue, Jiang, Zhao, and Ma (2017) suggested a self-adaption mechanism in the ABC algorithm and used a global best solution to improve its exploitation capabilities. Adaption is required to fine tune control parameters during execution per the environmental changes. Xue et al. make use of flags to identify the successful and unsuccessful evolutions. Successful evaluations are used to improve solutions. The K-means clustering also improved using proposed approach. Tiwari and Kumar (2016) introduced a weight-driven approach in ABC with the logic that the solution with the high fitness value has more possibility of selection for next iteration, whereas a solution is likely to be discarded if it has low fitness. This strategy is similar to fitness-based position updating in ABC (Kumar, Kumar, Dhayal, & Swetank, 2014). Bansal, Jadon, Tiwari, Kiran, and Panigrahi (2017) used global and local search in ABC to solve the optimal power flow problem. The introduction of global and local search strategies in ABC improves the balancing between exploration and exploitation. The proposed modifications in the position update equation are as shown in Equation (6.7).

$$L_{ij} = x_{ij} + (prob_i) \times (x_{lj} - x_{ij}) + \phi_{ij} \times (x_{r1j} - x_{r2j}) \qquad (6.5)$$

$$G_{ij} = x_{ij} + (1 - prob_i) \times (x_{gj} - x_{ij}) + \phi_{ij} \times (x_{R1j} - x_{R2j}) \qquad (6.6)$$

$$v_{ij} = (prob_i) \times G_{ij} + (1 - prob_i) \times L_{ij} \qquad (6.7)$$

Where Equations (6.5) and (6.6) represent the local (L_{ij}) and global (G_{ij}) components, respectively. $r_1 \neq r_2 \neq i$ and $R_1 \neq R_2 \neq i$. ϕ_{ij} denotes an arbitrary number in [-1,1], and $prob_i$ denotes the probability of the ith solution that is decided by fitness of individual.

Sharma, Gupta, and Sharma (2016) incorporated fully informed learning (FIL) at the onlooker bee stage in ABC. Every individual grabs information about the best solutions in the current swarm and in its proximity. It assumes that best solutions update their position according to information collected in their neighbourhood in order to improve their exploitation capability. The position update equation in FIL is:

$$v_{ij} = x_{ij} + \phi_{ij} \frac{\sum_{k=1}^{SN}(x_{ij} - x_{kj})}{SN} + \psi_{ij}(y_j - x_{ij}) \qquad (6.8)$$

Where ϕ_{ij} and ψ_{ij} are uniformly distributed random numbers in the range [-1, 1] and [0, C], respectively. C is a positive constant. y_j denotes global best solution in jth dimension.

Akay and Karaboga (2017) anticipated nine variants of ABC for constrained optimization problems. The proposed variants modified the local search process, each parameter being taken from different solutions, the use of neighbourhood topology, the use of mutation and crossover operators from DE, and the use of global best and adaption in the scout bee phase. With the help of these experiments, it is concluded that performance of the ABC varies with the methods of the constraint-handling process.

An attractive force model in ABC is based on the gravity model proposed by Xiang et al. (2018). The newly proposed model improves the exploitation capability of ABC because, with the help of the attractive force model, better neighbours are selected in the proximity of current solutions. A global information interaction system was implemented for the onlooker bees with two novel search equations and an adaptive selection mechanism by Lin et al. (2018). A variable search strategy was incorporated in ABC by Kiran, Hakli, Gunduz, and Uguz (2015) for continuous optimization, and they proposed five different update rules. The problems considered have characteristics, and with the help of these modifications, they achieved better performance for the ABC. Karaboga and Kaya (2016) used a new adaptivity coefficient in the ABC and also used crossover rate to get a good rate of convergence. It was applied for the training of adaptive neuro-fuzzy inference system (ANFIS). The proposed modification gives better results in comparison to the popular neuro-fuzzy-based model. Sharma, Sharma, Sharma, and Bansal (2016) incorporated a disruption operator in the ABC algorithm. The newly introduced operator fluctuation in the swarm tries to balance the process of exploration and exploitation. Hussain, Gupta, Singh, Trivedi, and Sharma (2015) used a shrinking hypersphere local search strategy in the ABC and a position update strategy modified as follows:

$$x_{new,j} = x_{best,j} + \phi \times (x_{best,j} - r_j) \qquad (6.9)$$

Where $x_{best,j}$ is best solution in the current swarm, and r denotes a radius of hyperspace with x_{best} as center. It is initialized with a Euclidian distance and then updated using Equation (6.10).

$$r_j = r_j - \left\{ \frac{x_{best,j} - x_{worst,j}}{LSI} \right\} \qquad (6.10)$$

The ABC algorithm is hybridized to solve job-shop-scheduling problems without waiting by Sundar, Suganthan, and Jin et al. (2017), and Sharma and Pant (2017) presented a hybrid of ABC with shuffled frog leaping algorithm in order to improve balancing between exploration and exploitation. A gene recombination operator was constructed by combining different genes of superior quality in order to get better offspring by Li et al. (2017) and implanted with nine different variants of ABC. Karaboga and Akay performed a comparative study of the ABC (Karaboga & Akay, 2009), and Karaboga and Basturk analyzed the performance of the ABC (Karaboga & Basturk, 2008) for various problems. It is concluded that ABC is a better choice in comparison to other competitive nature-inspired algorithms. The performance of ABC can be enhanced by hybridizing it with other local search strategies because ABC is cooperative with other algorithms.

6.2.5 Popular Variants of the Artificial Bee Colony Algorithm

The ABC algorithm has been modified a number of times to get better performance for various optimization problems. A few of examples are listed here.

6.2.5.1 The Best-So-Far Artificial Bee Colony Algorithm

Banharnsakun, Achalakul, and Sirinaovakul (2011) wished for a variant of the ABC that shared the information about the best solutions found so far with the current population because this accelerates the process of finding the global best solutions. Additionally, the radius of the search space would be adjusted dynamically; it would be large for new solutions and get reduced with increasing iterations. The proposed strategy was named "best-so-far ABC" (BSFABC). The BSFABC suggested an objective value-based comparison to select best among current and new solutions. This method can be summarized as follows:

> *If (for minimization problems)*
> *If $(f_{new}(x) < f_{old}(x))$*
> * Select the new solution and discard old solutions*
> *Else*
> * Keep the old solution*
> *Else If (for maximization problems)*
> *If $(f_{new}(x) > f_{old}(x))$*
> * Select the new solution and discard old solutions*
> *Else*
> * Keep the old solution*

The BSFABC suggested changes in the search radius so that early iterations can explore the feasible search space and later iterations are able to exploit the best feasible solutions in the proximity of best solutions found so far. The new method also gets rid of trapping into local optima, the modified position update equation is shown in Equation (6.11).

$$v_{ij} = x_{ij} + \phi_{ij}\left[\omega_{max} - \frac{iteration}{MCN}(\omega_{max} - \omega_{min})\right]x_{ij} \qquad (6.11)$$

Where MCN denotes maximum cycle number, ω_{max}, and ω_{min} denotes maximum and minimum percentage of position adjustment.

Furthermore, this algorithm shares the information about best solution in the current solutions and concentrates its searching process near the best solutions only, to increase speed of convergence, because it is more probable that the optimum solution lies in proximity of best solutions found so far. The BSFABC is a very popular successful variant of the ABC and is used to solve numerous real-world problems. A critical analysis of performance and sensitivity of various variants of ABC is carried out in (Banharnsakun, Sirinaovakul, & Achalakul, 2012a) and the results compared with BSFABC. The inventors of BSFABC applied it to solve an NP-complete problem, namely the job-shop-scheduling problem (Banharnsakun, Sirinaovakul, & Achalakul, 2012b). In (Banharnsakun et al., 2012b), a novel local search strategy, named "variable neighbourhood search" is applied on the best solution found so far by BSFABC after the scout bee phase. In (Banharnsakun, Sirinaovakul, & Achalakul, 2013) BSFABC with multiple patrilines is implemented to get rid of clustering problems. The BSFABC also proved its performance for template matching-based object detection (Banharnsakun & Tanathong, 2014) with a better rate of convergence and greater accuracy. Nantapat, Kaewkamnerdpong, Achalakul, and Sirinaovakul (2011) designed a nano-robot swarm with the help of the best-so-far ABC algorithm that has applications in the medical field and a framework demonstrated through artificial platelets for primary hemostasis in order to avoid hemorrhage.

6.2.5.2 The Gbest-Guided Artificial Bee Colony Algorithm

Zhu and Kwong (2010) proposed an improved version of the ABC motivated by particle swarm optimization, named the "Gbest-guided ABC" (GABC) algorithm. It has been observed that the performance of the ABC depends on two contradictory processes, namely exploration and exploitation. Proper balancing between these two processes improves the rate of convergence. The position update equation in basic ABC (Karaboga, 2005) is driven by the current position and a randomly generated step size. In order to improve the process, a new position update equation was suggested in (Zhu & Kwong, 2010), as shown in Equation (6.12).

$$v_{ij} = x_{ij} + \phi_{ij} \times (x_{ij} - x_{kj}) + \psi_{ij} \times (y_j - x_{ij}) \qquad (6.12)$$

Where $\psi_{ij} \times (y_j - x_{ij})$ is newly added term and named as the gbest term. ψ_{ij} is a uniformly distributed random number in the range [0, C] for a positive constant C, and y_j denotes the global best solution in the current swarm. This equation improves the exploitation of best feasible solution as it tries to improve the global best solution in the current swarm.

Additionally, the GABC is modified a number of times and used to solve various problems. Garg, Jadon, Sharma, and Palwalia (2014) applied GABC to analyze a power system network and solved the problem of load flow with a five-bus system that had previously been solved by an inefficient classical technique, namely the Newton Raphson method. Jadhav and Roy (2013) used GABC for an economic load dispatch problem with inclusion of wind energy. In (Jadhav & Roy, 2013) the probability calculation method was modified for better exploitation capability, as shown below:

$$P_i = \frac{0.9 \times fit_i}{fit_{best}} + 0.1 \qquad (6.13)$$

Where fit_i and fit_{best} denote the fitness of *ith* solution and highest fitness, respectively.

A discrete gbest ABC algorithm was suggested in (Huo, Zhuang, Gu, Ni, & Xue, 2015) for optimal cloud service composition with a time attenuation function. A modified gbest ABC algorithm proposed in (Bhambu, Sharma, & Kumar, 2018) with two modifications in the basic GABC. It modified the employed bee phase and the onlooker bee phase and suggested a new position update process, as shown in Equations (6.14) and (6.15), respectively.

$$xnew_{ij} = x_{ij} + \phi_{ij}(x_{ij} - x_{kj}) + \frac{(1-t)}{T} \times \psi_{ij}(y_j - x_{ij}) \qquad (6.14)$$

$$xnew_{ij} = x_{ij} + \frac{(1-t)}{T} \times \phi_{ij}(x_{ij} - x_{kj}) + \psi_{ij}(y_j - x_{ij}) \qquad (6.15)$$

Where t and T denotes current iteration and total number of iterations, respectively. The rest of the symbols have their usual meaning. In Equations (6.14) and (6.15), there are two components: $\phi_{ij}(x_{ij} - x_{kj})$ and $\psi_{ij}(y_j - x_{ij})$. The first component, $\phi_{ij}(x_{ij} - x_{kj})$, maintains the stochastic nature in the proposed algorithm, and the second component, $\psi_{ij}(y_j - x_{ij})$, drives toward the global optimum and speeds up the convergence. The weight of these components changes iteratively to maintain the proper balance between intensification and diversification.

Sharma, Sharma, and Kumar (2016) recently anticipated a novel strategy in the ABC by considering the local best and global best solutions. The employed bee and onlooker bee phases are modified as shown in Equations (6.16) and (6.17), respectively.

$$x_{new,j} = x_{old,j} + \phi \times (x_{old,j} - x_{kj}) + \psi \times (x_{Lbest,j} - x_{old,j}) \tag{6.16}$$

$$x_{new,j} = x_{old,j} + \phi \times (x_{old,j} - x_{kj}) + \psi \times (x_{Gbest,j} - x_{old,j}) \tag{6.17}$$

Where $x_{Lbest,j}$ denotes the local best value in the current population, and $x_{Gbest,j}$ denotes the global best value in the current population; the rest of the symbols have their usual meaning.

6.2.5.3 Memetic Search in the Artificial Bee Colony Algorithm

Memetic algorithms are defined as "instructions for carrying out behaviour, stored in brains." These algorithms are implemented in population-based algorithms as a local search to improve the quality of solutions by exploring the search space. Bansal, Sharma, Arya, and Nagar (2013) introduced a new phase in the ABC as a local search phase, namely the memetic search phase, and proposed an algorithm named as memetic ABC (MeABC). In the newly introduced phase, step size is controlled by a golden section search (GSS) process. In case of the ABC, the step size depends on φ_{ij} as shown in Equation (6.2). The proposed strategy performs fine tuning of φ_{ij} dynamically in every iteration. The memetic search phase is summarized in Algorithm 1.

> **Algorithm 1: Golden section search**
> *Initialize a = -1.2 and b = 1.2*
> *Repeat while ($|a - b| < \varepsilon$)*
> *Compute* $F_1 = (b - (b - a) \times \psi)$, $F_2 = (a + (b - a) \times \psi)$
> *Calculate $f(F_1)$ and $f(F_2)$ for objective function*
> *if $f(F_1) < f(F_2)$ then*
> > $b = F_2$
>
> *else*
> > $a = F_1$

Using F_1 and F_2, it generates new solutions, and based on those solutions, the best feasible solution in current swarm is identified. The position is updated using the following equation.

$$x_{newj} = x_{bestj} + F \times (x_{bestj} - x_{kj}) \tag{6.18}$$

The MeABC algorithm has been modified by various researchers because it is very efficient. Kumar, Sharma, and Kumari (2014a, 2014b), Kumar, Kumar, Sharma, and Sharma (2014) proposed modifications in the MeABC to improve balancing between exploration and exploitation.

6.2.5.4 The Hooke-Jeeves Artificial Bee Colony Algorithm

The Hooke-Jeeves method is one of the popular pattern search methods. Kang, Li, Ma, and Li (2011) hybridized the ABC with a modified Hooke-Jeeves method, which takes advantage of previous moves and has an adaptable step length. This new method is combination of two moves: the exploratory and pattern moves. The exploratory move is responsible for exploration of the search space, and the pattern move exploits the direction decided on by the exploratory move. A new fitness calculation method was also suggested in (Kang et al., 2011) based on ranking of solutions, computed as follow:

$$fit_i = 2 - SP + \frac{2(SP - 1)(p_i - 1)}{NS - 1} \tag{6.19}$$

Where SP is a new parameter named "selection pressure." NS denotes the number of solutions or population size, and pi denotes the position of ith solution in the current population. The proposed variant outperformed basic ABC and the differential evolution algorithm.

6.2.5.5 The Multi-Objective Artificial Bee Colony Algorithm

The ABC algorithm performs very well for a single objective problem and also for multi-objective optimization problems. Multi-objective ABC was implemented with Pareto dominance theory in (Akbari, Hedayatzadeh, Ziarati, & Hassanizadeh, 2012). The flying behaviour of individuals is controlled by an external archive and a Pareto front in these archives, assessed by a grid-based approach. The fitness of individuals is assessed by Equation (6.20) in (Akbari et al., 2012).

$$fit(\vec{x_m}) = \frac{dom(m)}{FoodNumber} \tag{6.20}$$

Where $dom(m)$ denotes number of food sources dominated by food source m.

6.2.5.6 The Lévy Flight Artificial Bee Colony Algorithm

The Lévy flight search scheme was integrated with ABC in (Sharma, Bansal, Arya, & Yang, 2016). Lévy flight is a random walk in a multi-dimensional search space with a step length depending on the probability distribution. The main task of a Lévy flight search in a hybridized algorithm is to discover the feasible search space, and it is implemented as a local search strategy in the ABC, in addition to existing three phases.

The random step lengths are generated by the Lévy distribution shown in Equation (6.21):

$$L(s) \sim |s|^{-1-\beta} \tag{6.21}$$

Where β is an index with suggested range $(0 < \beta \leq 2)$, and s denotes the size of step. The random step sizes are generated using some probabilistic distribution such as Mantega's algorithm. The step length s is computed using Equation (6.22).

$$s = \frac{u}{|v|^{\frac{1}{\beta}}} \tag{6.22}$$

Where $\underline{u}$ and v are computed using a normal distribution as shown in Equation (6.23).

$$u \sim N(0, \sigma_u^2), \ v \sim N(0, \sigma_v^2) \tag{6.23}$$

Where,

$$\sigma_u = \left\{ \frac{\Gamma(1+\beta)\sin(\pi\beta/2)}{\beta\,\Gamma[(1+\beta)/2]2^{(\beta-1)/2}} \right\}^{1/\beta}, \ \sigma_v = 1. \tag{6.24}$$

Where $\Gamma(.)$ denotes a gamma function and is computed using Equation (6.25).

$$\Gamma(1+\beta) = \int_0^\infty t^\beta e^{-t} dt \tag{6.25}$$

If β is an integer, then $\Gamma(1+\beta) = \beta!$.

In (Sharma et al., 2016), the step sizes are derived from the Lévy distribution to exploit the search area and computed using Equation (6.26)

$$step\_size(t) = 0.001 \times s(t) \times SLC \tag{6.26}$$

Where t represents the iteration counter for local search strategy, and s(t) is decided by using Equation (6.22) and the social learning component of the global search algorithm represented by SLC. A multiplier 0.001 is used Equation (6.26) to keep the step size in the boundary because it is very aggressive.

6.2.6 Real-World Applications of the Artificial Bee Colony Algorithm

The ABC algorithm is one of the most efficient stochastic algorithms in the field of optimization. Since its inception, the popularity of ABC has been increasing day by day. Initially, it was developed for unconstrained optimization problems and then applied for constrained optimization problems. The application list of ABC is very large, including feature selection and attribute identification (Hancer et al., 2018), K-means based clustering problem (Xue et al., 2017), constrained and unconstrained engineering optimization problems (Akay & Karaboga, 2017; Akbari et al., 2012; Bansal et al., 2013; Bhambu et al., 2018; Hussain et al., 2015; Kang et al., 2011; Kiran et al., 2015; Kumar, Kumar, Dhayal, & Swetank, 2014; Kumar, Sharma, & Kumari, 2014b; Li et al., 2017; Lin et al., 2018; Sharma et al., 2016, 2016, 2016; Tiwari & Kumar, 2016; Xiang et al., 2018; Zhu & Kwong, 2010), Optimal power flow problems (Bansal et al., 2017), ANFIS training (Karaboga & Kaya, 2016), job-shop scheduling with no-wait constraint (Banharnsakun et al., 2012b; Sundar et al., 2017), chemical engineering problems (Sharma & Pant, 2017), image registration problems (Banharnsakun et al., 2011), Template matching (Banharnsakun & Tanathong, 2014), Nano-robot swarms (Nantapat et al., 2011), load-flow problems (Zhu & Kwong, 2010), economic load dispatch problems (Jadhav & Roy, 2013), cloud service (Huo et al., 2015), and slop stability analysis (Kang et al., 2011). It has applications in almost every field, including computer science; engineering; mathematics; energy; agricultural sciences; biological sciences; physics; astronomy; material sciences; biochemistry; genetics; molecular biology; environmental sciences; business, management, and accounting; social sciences; and medical sciences among others.

One of the important areas of application of the ABC algorithm is neural network training. Karaboga, Akay, and Ozturk (2007) used the ABC algorithm to find an optimal weight set by training a feed-forward neural network. Karaboga and Akay (2007) established the ABC on training on artificial neural networks, which has applications in the area of signal processing. In 2009, the ABC algorithm was applied on training feed-forward neural networks to categorize diverse data sets by Karaboga and Ozturk (2009) to help the machine learning community. The ABC was employed to train a multilayer perceptron neural network for the purpose of classification of acoustic emission signals to their respective source by Omkar and Senthilnath (2009). Garro, Sossa, and Vázquez (2011) synthesized ANN connections among neurons using the ABC algorithm.

ABC has been used by some researchers to get rid of optimization problems in the field of engineering. Adaryani and Karami (2013) used the ABC algorithm to solve a multi-objective optimal power flow (OPF) problem and employed three test power systems. The problem is highly non-linear non-convex problem with multiple constraints and multiple objectives. Ayan and Kılıç (2012) solved an optimal reactive power flow problem by ABC and minimized active power loss in the power system. Abu-Mouti and El-Hawary (Abu-Mouti & El-Hawary, 2011) employed ABC to identify optimal distributed generation allocation and sizing in distribution systems. Khorsandi, Hosseinian, and Ghazanfari (2013) modified ABC to solve discrete and continuous OPF problems. The proposed modification includes a fuzzified multi-objective version of ABC. Rao and Pawar (2009) modeled and optimized the parameters of wire electrical discharge machining using ABC. Rao and Pawar (2010) used ABC for the parameter optimization of a multi-pass milling process and compared its results with PSO and SA.

ABC has various applications in the area of wireless sensor network (WSN) and mobile Adhoc networks also. An energy-efficient clustering approach was developed by Karaboga, Okdem, and Ozturk (2012) using ABC to increase the lifetime of a network. Ozturk, Karaboga, and Gorkemli (2011) deployed WSN dynamically using ABC and got better performance in comparison to stationary sensors. Udgata, Sabat, and Mini (2009) deployed sensors (to establish a WSN) in irregular terrain to maximize the covered area and to minimize cost by deploying fewer sensors using the ABC algorithm.

The ABC algorithm also has a strong influence in the area of image processing. Horng (2011) applied ABC for image segmentation using multi-level maximum entropy thresholding. The results were compared with four popular population-based algorithms and proved that ABC is the best choice to select the threshold for segmenting an image. Ma, Liang, Guo, Fan, and Yin (2011) employed ABC for segmentation synthetic aperture radar. Sharma, Arya, and Saraswat (2014) used ABC on tissue images for the purpose of segmentation of leukocytes. The proposed strategy automatically performs segmentation of leukocytes. ABC has application in almost every area of optimization. A detailed study on the application of ABC was carried out by Karaboga, Gorkemli, Ozturk, and Karaboga (2014) and Bansal, Sharma, and Jadon (2013).

6.3 The Firefly Algorithm

6.3.1 Introduction

The firefly has a variety of species and is mostly found in tropical regions. Most of the fireflies produce rhythmic flashing for a short time. This flashing is due to a chemical reaction in living entities termed bioluminescence. The pattern of flashing varies from species to species of firefly. The firefly algorithm (FA) is based on the flashing behaviour of fireflies. It is an intelligent stochastic algorithm for global optimization problems, invented by Xin-She Yang (2009). The glow of the firefly is created in the lower abdomen through a chemical reaction identified as bioluminescence. This bioluminescence is an essential part of mating ritual in fireflies and is the prime medium of communication between male and female fireflies. This light is also used to entice a mating partner or prey. This flashing also helpful in protecting fireflies' territory and warning predators away for a safe habitat. In FA, it is assumed that all fireflies are unisex and attracted toward each other regardless of their sex. This algorithm is established on the basis of two key concepts: the strength of the light emanated and the degree of attractiveness that is engendered between two fireflies.

6.3.2 The Behaviour of Natural Fireflies

The natural firefly demonstrates amazing flashing behaviour while searching for prey, attracting mates, and protecting its territory. Mostly fireflies live in a tropical environment. Generally, they produce cold light such as green, yellow, or pale-red light. The attractiveness of a firefly depends on its illumination intensity, and for any pair of fireflies, the brighter one will attract the other; so, the less bright one is moved toward the brighter one. The brightness of light decreases with increasing distance. The flashing pattern may vary from one species to another species of firefly. In the case of some firefly species, the female hunts other species using this phenomenon. Some fireflies show synchronized flashing behaviour in a large group for attracting prey. The female fireflies observe the flashing produced by male fireflies from a stationary position. After spotting an interesting flashing, the female firefly responds with flashing, and in this way mating ritual takes place. Some female fireflies produce the flashing pattern of other species and attract male fireflies to trap them to devour.

6.3.3 The Firefly Algorithm – Introduction

The firefly algorithm simulates the natural phenomenon of fireflies. The real firefly naturally demonstrates a discrete pattern of flashing, while the firefly algorithm assumes that they are always glowing. To simulate this flashing behaviour of the firefly, Xin-She Yang formed three rules (Yang, 2009).

1. It is assumed that all fireflies are unisex; therefore, one may be attracted toward any of the other fireflies.
2. Brightness of a firefly decides the degree of attractiveness; a brighter firefly attracts a less bright firefly. It moves randomly if no firefly is brighter than the firefly being considered.
3. The optimum value of the function is proportionate to the brightness of a firefly.

The intensity of light (I) and distance from source (r) follow inverse square law; thus, the intensity of light (I) decreases with increasing distance from the light source. This is due to absorption by air. This phenomenon bounds the visibility of the firefly in a very limited radius.

$$I \propto \frac{1}{r^2} \tag{6.27}$$

Major steps while implementing the firefly algorithm to solve a problem are as follows:

In first step, the objective function $f(x_i)$ is initialized for ith solution. The light intensity for ith solution is computed using Equation (6.28).

$$I_r = I_0 \, e^{-\gamma r^2} \tag{6.28}$$

Where I_0 is light intensity at distance $r = 0$, and γ denotes an absorption coefficient. Sometimes a monotonically decreasing function is also used, as shown in Equation (6.29)

$$I_r = \frac{I_0}{1 + \gamma r^2} \tag{6.29}$$

The second step of the firefly algorithm is initialization of population

$$x_{t+1} = x_t + \beta_0 e^{-\gamma r^2} + \alpha \, \varepsilon \tag{6.30}$$

Where t denotes generation counter, x_t represents current position of an individual, the term $\beta_0 e^{-\gamma r^2}$ is for attraction, and the term $\alpha \varepsilon$ is for randomization. The next step computes the attractiveness for fireflies using Equation (6.31).

$$\beta = \beta_0 e^{(-\gamma \times r^2)} \tag{6.31}$$

Where β_0 represents attractiveness when $r = 0$.

In next step, identify the movement of firefly with low illumination toward a brighter one.

$$x_i^{t+1} = x_i^t + \beta_0 e^{-\gamma r_{ij}^2}(x_j^t - x_i^t) + \alpha \, \varepsilon_i^t \tag{6.32}$$

In final stage, update the light intensities and assign a ranking to all fireflies in order to decide the current best solution. Figure 6.2 depicts major steps of firefly algorithm.

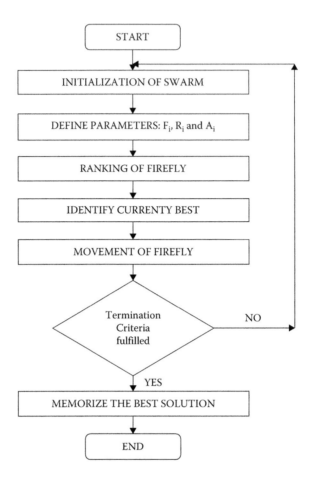

FIGURE 6.2
Flow chart for firefly algorithm.

The firefly algorithm has similarity with particle swarm optimization (PSO) and the bacterial foraging algorithm (BFA). Both FA and PSO have two major components in the position update equation; one is deterministic and the other is stochastic. In case of FA, the attractiveness is decided by two components: an objective function and distance, and in BFA the attraction among bacteria also has two components: fitness and distance. When the firefly algorithm is implemented, it takes two inner loops for whole population (say, n) and one outer loop for specified iterations (say, I). Thus, the computational complexity of FA is $O(n^2I)$ in the worst case.

6.3.4 Variants of the Firefly Algorithm

The firefly algorithm is only one decade old but is gaining popularity among researchers and scientists working in the field of optimization due to its simplicity and ease of implementation. Francisco, Costa, and Rocha (2014) performed some experiments on the firefly algorithm and used a mathematical function "norm." The norm is a function that firmly allocates a non-negative length. In (Francisco et al., 2014), a couple of new approaches were proposed for computation of attractiveness function. First, is computation of distance between two fireflies using p-norm. The second approach proposed two new attractiveness functions β_1 and β_2 as shown below.

$$\beta_{ij}^1 = \begin{cases} \beta_0 e^{(-\gamma \times r_{ij}^2)} & \text{if } r_{ij} \le d \\ 0.5 & \text{if } r_{ij} > d \end{cases} \tag{6.33}$$

$$\beta_{ij}^2 = \begin{cases} \beta_0 e^{\left(-\dfrac{f(x_i)-f(x_j)}{r_{ij}}\right)} & \text{if } f(x_i) > f(x_j) \\ 0 & \text{otherwise} \end{cases} \tag{6.34}$$

Where all the symbols have their usual meaning.

Yang et al. solved stochastic test problems and non-linear complex design problems using the firefly algorithm (Yang, 2010b) and non-convex economic dispatch problems (Yang, Hosseini, & Gandomi, 2012) and performed a state-of-the-art survey on recent variants and applications of FA (Yang & He, 2013b). Gandomi, Yang, Talatahari, and Alavi (2013) incorporated chaos into the firefly algorithm to improve the global search kinesis. Chaotic optimization algorithms are more efficient because the chaos is non-repetitive and ergodic in nature, whereas a random variable is not as efficient. In (Gandomi et al., 2013) authors used non-invertible one-dimensional maps to engender chaotic sets. Kaabeche et al. (Kaabeche, Diaf, & Ibtiouen, 2017) solved an economic load dispatch problem that considers solar and wind energy with load dissatisfaction problem and cost.

Wang, Wang, Zhou, et al. (2017) developed a new variant of FA based on the neighbourhood attraction model in place of the full attraction model. The original FA (Yang, 2009) is based on attraction among fireflies; the firefly with less brightness is attracted toward a brighter one. In this model, total number of attractions is $N(N-1)/2$, if there are N fireflies. This large number of attractions (the average number of attractions is $(N-1)/2$) leads to a high fluctuation rate while searching for optimality. This higher fluctuation rate leads to a slow convergence. If the number of fluctuations is very much less, then there are chances for premature convergence. In order to balance this fluctuation rate, a neighbourhood-based attraction model was proposed in (Wang et al., 2017a), inspired by the *k*-neighbour-hood concept. In this model, *k* fireflies are considered brighter than X_i, and fireflies are arranged in a circular topology (X_i connected with X_{i-1} and X_{i+1}). In this topology, the total number of attractions is $N*k$, that is considerably less in comparison to $N(N-1)/2$ because *k* is smaller than $(--1)/2$.

Wang et al., 2017c) proposed adaptive parameters in FA that modify its value according to iteration number, as shown in Equations (6.35) and (6.36).

$$\alpha(t+1) = \left(1 - \frac{t}{G_{max}}\right) \times \alpha(t) \tag{6.35}$$

$$\alpha(t+1) = \left(1 - \frac{Feval}{Max\_Feval}\right) \times \alpha(t) \tag{6.36}$$

Where *t* denotes the iteration counter, and the initial value of α is taken as 0.5. The first equation stops with G_{max}, whereas the second one stops with the maximum number of evaluations. Wang et al. also suggested change in attractiveness coefficient β as shown in Equation (6.37).

$$\beta_0(t+1) = \begin{cases} rand_1, & if \ rand_2 < 0.5 \\ \beta_0(t) & else \end{cases} \tag{6.37}$$

Where $rand_1$ and $rand_2$ are uniformly generated random numbers.

Chuah, Wong, and Hassan (2017) developed a new variant of FA by mixing three different strategies in firefly, and named it swap-based discrete FA (SDFA), which is applied to get rid of the most difficult problem, namely the travelling salesman problem. The new strategy integrated the firefly algorithm with reset strategy, nearest neighbour initialization, and fixed radius nearest neighbour. Here, the distance between two fireflies is computed using a swap distance strategy as follows:

$$r_{ij} = \frac{d_{swap}}{N-1} \times 10 \tag{6.38}$$

Where d_{swap} denotes the swap distance in the range [0, N], and N represents the number of cities. This implies that increasing the distance between cities leads to a gradual decrease in attractiveness β_{ij}. A nearest neighbour strategy was incorporated in this new variant that randomly starts with a city and keeps selecting neighbouring cities until the cycle is completed. Additionally, a reset approach is used to skip the local optimum. After generation of final travel path, a local optimization approach, namely fixed radius near neighbour 2-opt, is implemented in addition to SDFA (Chuah et al., 2017).

A multi-objective version of FA was developed by Wang et al., 2017b) using a crossover strategy in FA. The new solutions in the proposed algorithm are generated as follows:

$$x_{id}^*(t+1) = \begin{cases} x_{id}(t+1) & \text{if } rand_d\,(0,1) \leq CR \vee d = d_{rand} \\ x_{id}(t) & \text{otherwise} \end{cases} \tag{6.39}$$

Where t is the iteration counter, CR denotes crossover rate, $rand_d(0.1)$ is an arbitrary number in the range [0,1] and d_{rand} is an arbitrary number in the range [0, D] for a D-dimensional problem. The parameter α updated using Equations (6.35) and (6.36). The modified FA is used to solve a big data problem with thousands of decision variables.

6.3.5 Real-World Applications of the Firefly Algorithm

The application areas of FA are widespread, ranging from computer science, engineering, and mathematics to biochemistry, genetics, and molecular biology. It is equally useful in almost every field where one needs optimization. A detailed study on recent developments in FA and its applications is provided in (Yang & He, 2013b). Some of the applications of FA are: engineering design problems, feature selection, digital image compression, scheduling problems, load dispatch problems, the travelling salesman problem, classifications and clustering, the training of neural networks, and graph colouring among others.

Application of Firefly varies from the design of complex mechanical tools to managing power flow. Yang (2010b) applied the firefly algorithm on a complex design problem, namely pressure vessel design, and found an optimum result. A non-convex economic dispatch problem was solved by Yang et al. (2012) using FA and identified more economical loads in comparison to earlier methods. Kaabeche et al. (Kaabeche et al., 2017) engendered a variant of the firefly algorithm to find the optimal size of a hybrid renewable energy system by accounting for the load dissatisfaction rate (LDR) principles and the electricity cost (EC) pointer and attained the anticipated LDR with the lowest EC in comparison to another swarm-based optimization algorithm. Chuah et al. (2017) introduced a discrete swap-based variant of FA to solve the travelling salesman problem (TSP). TSP is an NP-hard problem with large search space.

The firefly algorithm gives excellent performance for clustering and classification problems. Senthilnath, Omkar, and Mani (2011) implemented FA for an unsupervised classification method, i.e., clustering. Clustering has various applications in image processing, statistical data investigation, data mining, and other fields of science, engineering, and management.

A wide-ranging study on the performance is provided by comparing FA with 11 diverse algorithms and establishing that FA is one of the best choices for efficient clustering. The FA is very successful in solving complex optimization problems with different complexity levels; that's why it has a large pool of applications.

6.4 The Bat Algorithm

6.4.1 Introduction

The bat algorithm (BA) is a swarm-intelligence-based algorithm inspired by the echolocation behaviour of micro-bats (Yang, 2010a) developed by Xin-She Yang in 2010. Most of the micro-bats radiate sounds out to the surroundings and listen to the echoes of those sounds that reappear from different objects to identify prey, elude obstacles, and trace their dark nestling cracks. The sound pulses vary with species of bats. Basically, frequency tuning acts as a mutation because it causes fluctuations in the solutions, principally in the vicinity of better solutions, although a large size of mutation leads to a global search. Particular selection is carried out by applying pressure for selection that is comparatively constant due to the use of the best solution in the present swarm that has been established up to now. In comparison to genetic algorithms, there is no obvious crossover; nevertheless, the deviations of loudness and pulse emission lead to a variation in mutation. Alternatively, there is the capability of auto-zooming in the sense that exploitation becomes concentrated as the search is approaching global optimality in the variations of loudness and pulse emission rates. This results in automatically switching to an exploitative phase from an explorative phase.

6.4.2 Overview of the Natural Behaviour of Bats

Bats are the only mammals that have wings, and they have the extraordinary ability of echolocation. They are the second largest mammals in terms of population with more than 1200 species. Generally they are divided into two classes: the echolocating micro-bats and the fruit-eating mega-bats. The bat algorithm was developed by Xin-She Yang (2010a) based on the behaviour of the first category of bats. Most of the bats rest in roosting posture, hanging upside down. All the micro-bats and some of the mega-bats emit ultrasonic sounds to produce echoes. The brain and acoustic nervous system of a micro-bat can generate an in-depth picture of the environment by comparing the outbound pulse with the recurring echoes. Micro-bats emit these ultrasounds (generated using the larynx) generally through the mouth and occasionally through the nose. They finish their call before the echoes return. The echolocation may be a low-duty cycle or a heavy-duty cycle. In the first case, bats can separate their calls and recurring echoes by time, and in the second case, bats emit an uninterrupted call and isolated pulse and echo in frequency. The echolocation is also known as bio-sonar. Bats radiate sounds out to their surroundings and listen to the echoes of those sounds that reappear from different objects around them. Echolocation is mainly used for the purpose of navigation and foraging by animals. With the help of these echoes, the bats measure size and distance of objects; some species of bats are even able to measure how fast an object is moving.

6.4.3 Mathematical Formulation of the Bat Algorithm

A bat flies randomly with velocity v_i at location x_i with a static frequency f_{min}, capricious wavelength λ, and loudness A_0 in search of target/object/prey. The frequency varies from f_{min} to f_{max}. The loudness of the sound may vary according to the requirement from A_0 to A_{min}. Xin-She Yang (2010a) established a set of rules to update velocity, position, and loudness for a bat while searching for prey. The mathematical formulation of the bat algorithm is shown below:

$$f_i = f_{min} + (f_{max} - f_{min}) \times \beta \tag{6.40}$$

$$v_i^{t+1} = v_i^t + (x_i^t - x_*) \times f_i \tag{6.41}$$

$$x_i^{t+1} = x_i^t + v_i^t \tag{6.42}$$

$$A_i^{t+1} = \alpha A_i^t, r_i^{t+1} = r_i^0[1 - exp(-\gamma t)] \tag{6.43}$$

$$A_i^t \rightarrow 0, r_i^t \rightarrow r_i^U, as\ t \rightarrow \infty \tag{6.44}$$

Where β is a uniformly generated random number in the range [0, 1]. x_* denotes the global best solution in current swarm. α, and γ are constants such that $0 < \alpha < 1$ and $\gamma > 0$.

After the selection of global best solution, each local solution (x_{old}) in current swarm updates its location using Equation (6.45).

$$x_{new} = x_{old} + \varepsilon A^t \tag{6.45}$$

Where ε is an arbitrary number [−1, 1]

The BA combines the best characteristics of particle swarm optimization, genetic algorithms, and harmony search; thus, it provides good results in comparison to these algorithms (Yang, 2010a). Fine tuning of parameters α and γ affects the rate of convergence. Implementation of the BA is a little bit complicated, but it has proved its performance and established itself as a good nature-inspired meta-heuristic.

6.4.4 Variants of the Bat Algorithm

The bat algorithm has shown good results (Yang, 2010a) for optimization problems. Recently it has undergone numerous modifications, and some variants have been developed by hybridizing it with existing meta-heuristics or by introducing some new parameters. A few of them are discussed here.

6.4.4.1 Chaotic Bat Algorithm

Gandomi and Yang (2014) developed four variants of chaotic BA, and each variant was validated with the help of 13 chaotic maps. These chaotic maps replaced fixed parameters and enhanced flexibility and diversity because a different chaotic map leads to diverse behaviour by the algorithm. The proposed modifications are suggested changes in parameter β, λ, Loudness (A) and Pulse emission rate (r). The suggested changes in parameter β and λ are shown in Equations (6.46) and (6.47), respectively.

$$f_i = f_{min} + (f_{max} - f_{min}) \times CM_i \qquad (6.46)$$

$$v_i^{t+1} = v_i^t + (x_i^t - x_*) \times CM_i \times f_i \qquad (6.47)$$

In (Gandomi & Yang, 2014) Loudness (A) and Pulse emission rate (r) are also replaced by chaotic map.

6.4.4.2 Binary Bat Algorithm

Mirjalili, Mirjalili, and Yang (2014) engendered a binary version of BA after modifying some basic concepts of the velocity and position, modernizing the procedure. A transfer function has been introduced to update position in discrete binary search space because it is not possible to update position with existing method. The transfer function used is:

$$S(v_i^k(t)) = \frac{1}{1 + e^{-v_i^k(t)}} \qquad (6.48)$$

Where $v_i^k(t)$ denotes the velocity of ith individual in kth dimension. The proposed position update equation (position of ith individual in kth dimension) using a transfer function is:

$$x_i^k(t+1) = \begin{cases} 0 \; if \;\; rand < S(v_i^k(t+1)) \\ 1 \; if \;\; rand \geq S(v_i^k(t+1)) \end{cases} \tag{6.49}$$

In this process, each individual is assigned either 0 or 1 and due to this, the position of an individual remains unchanged with an increase in velocity. To overcome this issue, a new v-shaped transfer function was proposed and new position update process designed using this function, as shown in Equations (6.50) and (6.51), respectively.

$$V(v_i^k(t)) = \left| \frac{2}{\pi} arctan \left(\frac{\pi}{2} v_i^k(t) \right) \right| \tag{6.50}$$

$$x_i^k(t+1) = \begin{cases} (x_i^k(t))^{-1} \; if \;\; rand < V(v_i^k(t+1)) \\ \quad x_i^k(t) \;\;\; if \;\; rand \geq V(v_i^k(t+1)) \end{cases} \tag{6.51}$$

6.4.4.3 Parallel Bat Algorithm

The parallel version of BA was developed by Tsai, Dao, Yang, and Pan (2014). In parallel BA, the swarm is divided into some subgroups, and then these subgroups work independently. In (Tsai et al., 2014) the swarm is divided into G groups and the total iteration contains R ($R = \{R_1, 2R_1, 3R_1, ...\}$) times of communication. Parallel BA is summarized in Algorithm 2.

Algorithm 2: Parallel Bat Algorithm
Step 1: **Initialization:** Engender population of bats and construct G subgroups. Initialize each subgroup individually and define the iteration set R to implement the communication strategy.
Step 2. **Valuation:** Assess the value of each bat in every subgroup.
Step 3. **Update:** Apprise the velocity and position of the bat using basic BA (Yang, 2010a).
Step 4. **Communication Policy:** Select the best bat (G^t) in the current swarm and migrate in each group, and replace the worst bats with mutated G^t in each group in every R_1 iterations.
Step 5. **Termination:** Repeat step 2 to step 5 until the termination criteria are fulfilled. Memorize the best value of the function f (G^t) and the position of best bat among all the bats G^t.

6.4.4.4 Directional Bat Algorithm

Chakri, Khelif, Benouaret, and Yang (2017) came up with a novel variation of BA, namely directional BA (dBA) with four new amendments to basic

BA, to solve higher-dimensional problems. The bats produce sound pulses in two different directions, one in the direction of the best in the current swarm and another in a random direction and decide the position of the best feasible food source with the help of the fitness of individuals. The mathematical formulation of movement of bats given by Equation (6.52) (Chakri et al., 2017).

$$
\begin{cases}
x_i^{t+1} = x_i^t + (x^* - x_i^t)f_1 + (x_k^t - x_i^t)f_2 & \text{if } F(x_k^t) < F(x_i^t) \\
x_i^{t+1} = x_i^t + (x^* - x_i^t)f_1 & \text{otherwise}
\end{cases}
\tag{6.52}
$$

Where, x^* represents the best bat in the current swarm and x_k^t denotes a randomly selected bat in kth direction. F(.) denotes a fitness function that is essentially a function of the solution value. The frequencies of these two pulses are computed using Equation (6.53) with the help of two uniformly generated random number $rand_1$ and $rand_2$ in the range [0, 1].

$$
\begin{cases}
f_1 = f_{min} + (f_{max} - f_{min}) \times rand_1 \\
f_2 = f_{min} + (f_{max} - f_{min}) \times rand_2
\end{cases}
\tag{6.53}
$$

The proposed strategy enhances exploration competency exclusively in early iterations, which can avoid premature convergence. The position of the bats is updated using the following equation in (Chakri et al., 2017), which include loudness (A^t) and a monotonically decreasing function (w_i^t).

$$
x_i^{t+1} = x_i^t + <A^t> \varepsilon w_i^t
\tag{6.54}
$$

Where w_i is a scaling parameter to regulate the searching process that is computed with the help of Equation (6.55).

$$
w_i^t = \left(\frac{w_{i0} - w_{i\infty}}{1 - t_{max}} \right) \times (t - t_{max}) + w_{i\infty}
\tag{6.55}
$$

Initial (w_{i0}) and final ($w_{i\infty}$) value of scaling parameter computed using Equations (6.56) and (6.57), respectively.

$$
w_{i0} = (Ub_i - Lb_i)/4
\tag{6.56}
$$

$$
w_{i\infty} = w_{i0}/100
\tag{6.57}
$$

Chakri et al. (2017) suggested changes in pulse rate and loudness, as shown in Equations (6.58) and (6.59).

$$r^t = \left(\frac{r_0 - r_\infty}{1 - t_{max}}\right) \times (t - t_{max}) + r_\infty \tag{6.58}$$

$$A^t = \left(\frac{A_0 - A_\infty}{1 - t_{max}}\right) \times (t - t_{max}) + A_\infty \tag{6.59}$$

The proposed two sets of pulse emanations are emitted in diverse directions, which leads to a proficient algorithm. Exploration and exploitation are both improved because it leads the moves more onto the favorable search space in the directions of the best bats. Additionally, the proposed adaption in pulse emanation rates and the loudness control the exploration and exploitation at different stages of iterations.

6.4.4.5 Self-Adaptive Bat Algorithm

Fister, Fong, and Brest (2014) proposed a self-adaptive BA with the expectation that control parameters would be automatically adjusted during the algorithm's run. The suggested adaption in loudness and pulse rate are:

$$A^{t+1} = \begin{cases} A_{lb}^{(t)} + rand_0(A_{ub}^{(t)} - A_{lb}^{(t)}) & \text{if } rand_1 < \tau_1, \\ A^{(t)} & \text{otherwise} \end{cases} \tag{6.60}$$

$$r^{(t+1)} = \begin{cases} r_{lb}^{(t)} + rand_2(r_{ub}^{(t)} - r_{lb}^{(t)}) & \text{if } rand_3 < \tau_2, \\ r^{(t)} & \text{otherwise} \end{cases} \tag{6.61}$$

Where τ_i for $i = \{1,2\}$ denotes learning rate and fixed at $\tau_i = 0.1$, and *rand* denotes a random function that engenders an evenly distributed arbitrary number in the range [0, 1]. Using this self-adaptive strategy, every 10th candidate solution is re-formed. Further, Fister et al. (2014) hybridized this self-adaptive technique with one popular evolutionary algorithm, namely differential evolution (DE). The best characteristics of nature-inspired meta-heuristics is that they are cooperative with each other, not competitive, thus they can be hybridized with each other. The hybridized BA updates trial solutions using DE strategy as follows:

$$y_n = \begin{cases} DE\_Strategy, & \text{if } rand(0,1) \leq CR \vee n = D, \\ x_{in}^{(t)} & \text{otherwise} \end{cases} \tag{6.62}$$

The DE strategy that is considered the current best solution to generate trail solutions is considered for implementation. The four strategies considered are:

DE/rand/1/bin

$$y_j = x_{r1,j} + F \times (x_{r2,j} - x_{r3,j}) \qquad (6.63)$$

DE/best/1/bin

$$y_j = x_{best,j} + F \times (x_{r1,j} - x_{r2,j}) \qquad (6.64)$$

DE/best/2/bin

$$y_j = x_{best,j} + F \times (x_{r1,j} + x_{r2,j} - x_{r3,j} - x_{r4,j}) \qquad (6.65)$$

DE/rand_to_best/1/bin

$$y_j = x_{i,j} + F \times (best_j - x_{i,j}) - F \times (x_{r1,j} - x_{r2,j}) \qquad (6.66)$$

The DE strategies used are the best solution in the modification operations, that is, "rand_to_best/1/bin," "best/2/bin," and "best/1/bin," normally direct the simulated bats in the direction of the present best solution. Therefore, it is expected that the global best solutions are in the proximity of local best solutions.

6.4.5 Real-World Applications of the Bat Algorithm

The bat algorithm has a diverse area of application such as continuous and combinatorial optimization, image processing, clustering, engineering optimization, etc. The unique feature of BA is that it is stimulated by the echolocation behaviour of bats, and the performance of BA is comparatively better than other competitive algorithms considered (Yang & He, 2013a). It has fewer control parameters and is easy to implement. Some of the recent developments on BA were discussed in previous sections. In (Mirjalili et al., 2014), BA is applied on various benchmark problems with diverse characteristics such as uni-model and multi-model. Mirjalili et al. (2014) solved optical buffer design problems from optical engineering. Photonic crystal (PC) is very popular due to its extensive variety of applications because the photonic crystal waveguide (PCW) has applications in time-domain signal processing, non-linear optics, and all-optical buffers. The new variant of BA was successfully used for PCW design with better efficiency. The BA also has been applied to combinatorial optimization and scheduling problems, parameter estimation, classifications, data mining, clustering, and image processing, and in fuzzy systems to design various complex machinery.

Dao, Pan, and Pan (2018) implemented BA in a parallel environment to get rid of an NP-hard problem, namely a job-shop-scheduling problem.

The BA has application in classification, clustering, and data mining as well. Recently Khan, Nikov, and Sahai (2011) came up with a study of a clustering problem for office workplaces by means of a fuzzy bat algorithm. According to level of ergonomic risk, clusters of workplaces were divided into three classes: low, moderate, and high risk, and this approach produced better performance than fuzzy C-means clustering. A state-of-the-art relative study of BA and other algorithms was presented by Khan and Sahari (2012a) in the context of e-learning while training feed-forward neural networks. A new variant of BA for clustering problems has been presented by Khan and Sahari (2012b) and named bi-sonar optimization.

The bat algorithm was used by Marichelvam and Prabaharam (2012) to solve hybrid-flow shop-scheduling problems in order to minimize the mean flow time and make-span. It is supported by the results that BA emerges as a sole choice for solving hybrid flow-shop-scheduling problems. In the field of image processing, Abdel-Rahman, Ahmad, and Akhtar (2012) proposed an alternate to BA to estimate the complete-body human pose, and they established that BA does this better than another competitive probabilistic algorithm. A variation of the bat algorithm with a mutation for image matching has been created by Zhang and Wang (2012), and they specified that their bat-based model is highly efficient and realistic in image matching in comparison to other classical and probabilistic models.

6.5 Conclusion

The artificial bee colony algorithm, firefly algorithm, and bat algorithm are discussed in detail in this chapter with recent modifications and applications in the real world. These algorithms are getting the attention of researchers very rapidly, while they are comparatively new in the area of swarm intelligence. This is happening because of their ease of implementation and simplicity. Population-based nature-inspired algorithms have applications in nearly every single field, including computer science; engineering; mathematics; energy; agricultural sciences; biological sciences; physics; astronomy; biochemistry; molecular biology; genetics; material sciences; environmental sciences; business, management and accounting; social sciences; and medical sciences, among others.

References

Abdel-Rahman, E. M., Ahmad, A. R., & Akhtar, S., (2012). A metaheuristic bat inspired algorithm for full body human pose estimation. In *Ninth Conference on Computer and Robot Vision* (pp. 369–375).

Abu-Mouti, F. S., & El-Hawary, M. E. (2011). Optimal distributed generation allocation and sizing in distribution systems via artificial bee colony algorithm. *IEEE Transactions on Power Delivery, 26*(4), 2090–2101.

Adaryani, M. R., & Karami, A. (2013). Artificial bee colony algorithm for solving multi-objective optimal power flow problem. *International Journal of Electrical Power & Energy Systems, 53*, 219–230.

Akay, B., & Karaboga, D. (2017). Artificial bee colony algorithm variants on constrained optimization. *An International Journal of Optimization and Control, 7*(1), 98–111.

Akbari, R., Hedayatzadeh, R., Ziarati, K., & Hassanizadeh, B. (2012). A multi-objective artificial bee colony algorithm. *Swarm and Evolutionary Computation, 2*, 39–52.

Ayan, K., & Kılıç, U. (2012). Artificial bee colony algorithm solution for optimal reactive power flow. *Applied Soft Computing, 12*(5), 1477–1482.

Banharnsakun, A., Achalakul, T., & Sirinaovakul, B. (2011). The best-so-far selection in artificial bee colony algorithm. *Applied Soft Computing, 11*(2), 2888–2901.

Banharnsakun, A., Sirinaovakul, B., & Achalakul, T. (2012a, December). The performance and sensitivity of the parameters setting on the best-so-far ABC. In *Asia-Pacific Conference on Simulated Evolution and Learning* (pp. 248–257). Springer.

Banharnsakun, A., Sirinaovakul, B., & Achalakul, T. (2012b). Job shop scheduling with the best-so-far ABC. *Engineering Applications of Artificial Intelligence, 25*(3), 583–593.

Banharnsakun, A., Sirinaovakul, B., & Achalakul, T. (2013). The best-so-far ABC with multiple patrilines for clustering problems. *Neuro Computing, 116*, 355–366.

Banharnsakun, A., & Tanathong, S. (2014). Object detection based on template matching through use of best-so-far ABC. *Computational Intelligence and Neuroscience, 2014*, 7.

Bansal, J. C., Jadon, S. S., & Tiwari, R. R., Kiran, D., & Panigrahi, B. K. (2017). Optimal power flow using artificial bee colony algorithm with global and local neighbourhoods. *International Journal of System Assurance Engineering and Management, 8*(Suppl. 4), 2158. https://doi.org/10.1007/s13198-014-0321-7.

Bansal, J. C., Sharma, H., Arya, K. V., & Nagar, A. (2013). Memetic search in artificial bee colony algorithm. *Soft Computing, 17*(10), 1911–1928.

Bansal, J. C., Sharma, H., & Jadon, S. S. (2013). Artificial bee colony algorithm: A survey. *International Journal of Advanced Intelligence Paradigms, 5*(1–2), 123–159.

Bhambu, P., Sharma, S., & Kumar, S. (2018). Modified gbest artificial bee colony algorithm. In *Soft Computing: Theories and Applications* (pp. 665–677). Berlin: Springer.

Chakri, A., Khelif, R., Benouaret, M., & Yang, X. S. (2017). New directional bat algorithm for continuous optimization problems. *Expert Systems with Applications, 69*, 159–175.

Chuah, H. S., Wong, L. P., & Hassan, F. H. (2017, November). Swap-based discrete firefly algorithm for travelling salesman problem. In *International Workshop on Multi-Disciplinary Trends in Artificial Intelligence* (pp. 409–425). Cham: Springer.

Dao, T. K., Pan, T. S., & Pan, J. S. (2018). Parallel bat algorithm for optimizing makespan in job shop scheduling problems. *Journal of Intelligent Manufacturing, 29*(2), 451–462.

Fister, I., Fong, S., & Brest, J. (2014). A novel hybrid self-adaptive bat algorithm. *The Scientific World Journal*.

Francisco, R. B., Costa, M. F. P., & Rocha, A. M. A. (2014, June). Experiments with Firefly Algorithm. In *International Conference on Computational Science and Its Applications* (pp. 227–236). Cham: Springer.

Gandomi, A. H., & Yang, X. S. (2014). Chaotic bat algorithm. *Journal of Computational Science*, 5(2), 224–232.

Gandomi, A. H., Yang, X. S., Talatahari, S., & Alavi, A. H. (2013). Firefly algorithm with chaos. *Communications in Nonlinear Science and Numerical Simulation*, 18(1), 89–98.

Garg, N. K., Jadon, S. S., Sharma, H., & Palwalia, D. K. (2014). Gbest-artificial bee colony algorithm to solve load flow problem. In *Proceedings of the Third International Conference on Soft Computing for Problem Solving* (pp. 529–538). Springer.

Garro, B. A., Sossa, H., & Vázquez, R. A. (2011, June). Artificial neural network synthesis by means of artificial bee colony (abc) algorithm. In *2011 IEEE Congress on Evolutionary Computation (CEC)* (pp. 331–338). IEEE.

Hancer, E., Xue, B., Zhang, M., Karaboga, D., & Akay, B. (2018). Pareto front feature selection based on artificial bee colony optimization. *Information Sciences*, 422, 462–479.

Huo, Y., Zhuang, Y., Gu, J., Ni, S., & Xue, Y. (2015). Discrete gbest-guided artificial bee colony algorithm for cloud service composition. *Applied Intelligence*, 42(4), 661–678.

Horng, M. H. (2011). Multilevel thresholding selection based on the artificial bee colony algorithm for image segmentation. *Expert Systems with Applications*, 38 (11), 13785–13791.

Hussain, A., Gupta, S., Singh, R., Trivedi, P., & Sharma, H. (2015, September). Shrinking hyper-sphere based artificial bee colony algorithm. In *2015 International Conference on Computer, Communication and Control (IC4)* (pp. 1–6). IEEE.

Jadhav, H. T., & Roy, R. (2013). Gbest guided artificial bee colony algorithm for environmental/economic dispatch considering wind power. *Expert Systems with Applications*, 40(16), 6385–6399.

Kaabeche, A., Diaf, S., & Ibtiouen, R. (2017). Firefly-inspired algorithm for optimal sizing of renewable hybrid system considering reliability criteria. *Solar Energy*, 155, 727–738.

Kang, F., Li, J., Ma, Z., & Li, H. (2011). Artificial bee colony algorithm with local search for numerical optimization. *Journal of Software*, 6(3), 490–497.

Karaboga, D. (2005). *An Idea Based on Honey Bee Swarm for Numerical Optimization*. Technical report-tr06, vol. 200, Erciyes University, Turkey Engineering Faculty, Computer Engineering Department.

Karaboga, D., & Akay, B. (2007, June). Artificial bee colony (ABC) algorithm on training artificial neural networks. In *IEEE 15th Signal Processing and Communications Applications, 2007* (SIU 2007) (pp. 1–4). IEEE.

Karaboga, D., & Akay, B. (2009). A comparative study of artificial bee colony algorithm. *Applied Mathematics and Computation*, 214(1), 108–132.

Karaboga, D., Akay, B., & Ozturk, C. (2007, August). Artificial bee colony (ABC) optimization algorithm for training feed-forward neural networks. In *International Conference on Modeling Decisions For Artificial Intelligence* (pp. 318–329). Berlin: Springer.

Karaboga, D., & Basturk, B. (2008). On the performance of artificial bee colony (ABC) algorithm. *Applied Soft Computing*, 8(1), 687–697.

Karaboga, D., Gorkemli, B., Ozturk, C., & Karaboga, N. (2014). A comprehensive survey: Artificial bee colony (ABC) algorithm and applications. *Artificial Intelligence Review, 42*(1), 21–57.

Karaboga, D., & Kaya, E. (2016). An adaptive and hybrid artificial bee colony algorithm (aABC) for ANFIS training. *Applied Soft Computing, 49*, 423–436.

Karaboga, D., Okdem, S., & Ozturk, C. (2012). Cluster based wireless sensor network routing using artificial bee colony algorithm. *Wireless Networks, 18*(7), 847–860.

Karaboga, D., & Ozturk, C. (2009). Neural networks training by artificial bee colony algorithm on pattern classification. *Neural Network World, 19*(3), 279–292.

Khan, K., Nikov, A., & Sahari, A. (2011). A fuzzy bat clustering method for ergonomic screening of office workplaces. *Advances in Intelligent and Soft Computing, 101*, 59–66.

Khan, K., & Sahari, A. (2012a). A comparison of BA, GA, PSO, BP and LM for training feed forward neural networks in e-learning context. *International Journal of Intelligent Systems and Applications, 4*(7), 23–29.

Khan, K., & Sahari, A. (2012b). A fuzzy c-means bi-sonar-based metaheuristic optimization algorithm. *International Journal of Interactive Multimedia and Artificial Intelligence, 1*(7), 26–32.

Khorsandi, A., Hosseinian, S. H., & Ghazanfari, A. (2013). Modified artificial bee colony algorithm based on fuzzy multi-objective technique for optimal power flow problem. *Electric Power Systems Research, 95*, 206–213.

Kiran, M. S., Hakli, H., Gunduz, M., & Uguz, H. (2015). Artificial bee colony algorithm with variable search strategy for continuous optimization. *Information Sciences, 300*, 140–157.

Kumar, A., Kumar, S., Dhayal, K., & Swetank, D. K. (2014). Fitness based position update in artificial bee colony algorithm. *International Journal of Engineering Research & Technology, 3*(5), 636–641.

Kumar, S., Kumar, A., Sharma, V. K., & Sharma, H. (2014b, August). A novel hybrid memetic search in artificial bee colony algorithm. In *2014 Seventh International Conference on Contemporary Computing (IC3)* (pp. 68–73). IEEE.

Kumar, S., Sharma, V. K., & Kumari, R. (2014a, May). Memetic search in artificial bee colony algorithm with fitness-based position update. In *Recent Advances and Innovations in Engineering (ICRAIE)* (pp. 1–6). IEEE.

Kumar, S., Sharma, V. K., & Kumari, R. (2014b). Randomized memetic artificial bee colony algorithm. *International Journal of Emerging Trends & Technology in Computer Science, 3*(1), pp. 52–62

Li, G., Cui, L., Fu, X., Wen, Z., Lu, N., & Lu, J. (2017). Artificial bee colony algorithm with gene recombination for numerical function optimization. *Applied Soft Computing, 52*, 146–159.

Lin, Q., Zhu, M., Li, G., Wang, W., Cui, L., Chen, J., & Lu, J. (2018). A novel artificial bee colony algorithm with local and global information interaction. *Applied Soft Computing, 62*, 702–735.

Ma, M., Liang, J., Guo, M., Fan, Y., & Yin, Y. (2011). SAR image segmentation based on artificial bee colony algorithm. *Applied Soft Computing, 11*(8), 5205–5214.

Marichelvam, M. K., & Prabaharam, T. (2012). A bat algorithm for realistic hybrid flowshop scheduling problems to minimize make span and mean flow time. *ICTACT Journal on Soft Computing, 3*(1), 428–433.

Mirjalili, S., Mirjalili, S. M., & Yang, X. S. (2014). Binary bat algorithm. *Neural Computing and Applications, 25*(3–4), 663–681.

Nantapat, T., Kaewkamnerdpong, B., Achalakul, T., & Sirinaovakul, B. (2011, August). Best-so-far ABC based nanorobot swarm. In *2011 International Conference on Intelligent Human-Machine Systems and Cybernetics (IHMSC)*, (Vol. 1, pp. 226–229). IEEE.

Omkar, S. N., & Senthilnath, J. (2009). Artificial bee colony for classification of acoustic emission signal source. *International Journal of Aerospace Innovations*, *1*(3), 129–143.

Ozturk, C., Karaboga, D., & Gorkemli, B. (2011). Probabilistic dynamic deployment of wireless sensor networks by artificial bee colony algorithm. *Sensors, 11*(6), 6056–6065.

Rao, R. V., & Pawar, P. J. (2009). Modelling and optimization of process parameters of wire electrical discharge machining. *Proceedings of the Institution of Mechanical Engineers, Part B: Journal of Engineering Manufacture, 223*(11), 1431–1440.

Rao, R. V., & Pawar, P. J. (2010). Parameter optimization of a multi-pass milling process using non-traditional optimization algorithms. *Appl Soft Computing, 10*(2), 445–456.

Senthilnath, J., Omkar, S. N., & Mani, V. (2011). Clustering using firefly algorithm: Performance study. *Swarm and Evolutionary Computation, 1*(3), 164–171.

Sharma, H., Arya, K. V., & Saraswat, M. (2014, December). Artificial bee colony algorithm for automatic leukocytes segmentation in histopathological images. In *2014 9th International Conference on Industrial and Information Systems (ICIIS)* (pp. 1–6). IEEE.

Sharma, H., Bansal, J. C., Arya, K. V., & Yang, X. S. (2016). Lévy flight artificial bee colony algorithm. *International Journal of Systems Science, 47*(11), 2652–2670.

Sharma, H., Sharma, S., & Kumar, S. (2016, September). Lbest Gbest artificial bee colony algorithm. In *2016 International Conference on Advances in Computing, Communications and Informatics (ICACCI)*, (pp. 893–898). IEEE.

Sharma, K., Gupta, P. C., & Sharma, H. (2016). Fully informed artificial bee colony algorithm. *Journal of Experimental & Theoretical Artificial Intelligence, 28*(1–2), 403–416.

Sharma, N., Sharma, H., Sharma, A., & Bansal, J. C. (2016). Modified artificial bee colony algorithm based on disruption operator. In *Proceedings of Fifth International Conference on Soft Computing for Problem Solving* (pp. 889–900). Springer.

Sharma, T. K., & Pant, M. (2017). Shuffled artificial bee colony algorithm. *Soft Computing, 21*, 6085. https://doi.org/10.1007/s00500-016-21662.

Sundar, S., Suganthan, P. N., Jin, C. T., et al. (2017). A hybrid artificial bee colony algorithm for the job-shop scheduling problem with no-wait constraint. *Soft Computing, 21*, 1193. https://doi.org/10.1007/s00500-015-1852-9.

Tiwari, P., & Kumar, S. (2016). Weight driven position update artificial bee colony algorithm. In *International Conference on Advances in Computing, Communication, & Automation (ICACCA)* (Fall) (pp. 1–6). IEEE.

Tsai, C. F., Dao, T. K., Yang, W. J., & Pan, T. S. (2014, June). Parallelized bat algorithm with a communication strategy. In *International Conference on Industrial, Engineering and Other Applications of Applied Intelligent Systems* (pp. 87–95). Cham: Springer.

Udgata, S. K., Sabat, S. L., & Mini, S. (2009, December). Sensor deployment in irregular terrain using artificial bee colony algorithm. In *World Congress on Nature & Biologically Inspired Computing (NaBIC 2009)* (pp. 1309–1314). IEEE.

Wang, H., Wang, W., Cui, L., Sun, H., Zhao, J., Wang, Y., & Xue, Y. (2017a). A hybrid multi-objective firefly algorithm for big data optimization. *Applied Soft Computing*. https://doi.org/10.1016/j.asoc.2017.06.029

Wang, H., Wang, W., Zhou, X., Sun, H., Zhao, J., Yu, X., & Cui, Z. (2017b). Firefly algorithm with neighbourhood attraction. *Information Sciences, 382*, 374–387.

Wang, H., Zhou, X., Sun, H., Yu, X., Zhao, J., Zhang, H., & Cui, L. (2017c). Firefly algorithm with adaptive control parameters. *Soft Computing, 21*(17), 5091–5102.

Xue, Y., Jiang, J., Zhao, B., & Ma, T. (2017). A self-adaptive artificial bee colony algorithm based on global best for global optimization. *Soft Computing, 22*(9), 2935–2952.

Yang, X. S. (2009, October). Firefly algorithms for multimodal optimization. In *International Symposium on Stochastic Algorithms* (pp. 169–178). Springer.

Yang, X. S. (2010a). A new metaheuristic bat-inspired algorithm. In *Nature Inspired Cooperative Strategies for Optimization (NICSO 2010)* (pp. 65–74). Berlin: Springer.

Yang, X. S. (2010b). Firefly algorithm, stochastic test functions and design optimization. *International Journal of Bio-Inspired Computation, 2*(2), 78–84.

Yang, X. S., & He, X. (2013a). Bat algorithm: Literature review and applications. *International Journal of Bio-Inspired Computation, 5*(3), 141–149.

Yang, X. S., & He, X. (2013b). Firefly algorithm: Recent advances and applications. *International Journal of Swarm Intelligence, 1*(1), 36–50.

Yang, X. S., Hosseini, S. S. S., & Gandomi, A. H. (2012). Firefly algorithm for solving non-convex economic dispatch problems with valve loading effect. *Applied Soft Computing, 12*(3), 1180–1186.

Zhang, J. W., & Wang, G. G. (2012). Image matching using a bat algorithm with mutation. In *Applied Mechanics and Materials* 203, 88–93. Trans Tech Publications. Zurich, Switzerland.

Zhu, G., & Kwong, S. (2010). Gbest-guided artificial bee colony algorithm for numerical function optimization. *Applied Mathematics and Computation, 217*(7), 3166–3173.

7

Cuckoo Search Algorithm, Glowworm Algorithm, WASP, and Fish Swarm Optimization

Akshi Kumar

CONTENTS

7.1 Introduction to Optimization

The term optimization refers to making the best and most efficient use of a given situation. An optimization problem, thus, is one where we have to find the best solution among many feasible ones. A feasible solution is regarded as one that can fulfill all the constraints of a given problem. The best solution, on the other hand, is one that maximizes or minimizes certain problem parameters. Although the terms best and efficient can refer to various parameters depending on the problem at hand, in most real-world applications we need to either minimize the cost, time, or some other limited resource, or maximize the efficiency of the system. Before moving to the details of various nature-based optimization problem solving techniques, let's discuss the details regarding how a given optimization problem is formulated.

7.1.1 Formulation of an Optimization Problem

A general way to formulate an optimization problem mathematically is:

$$\underset{x \in R^n}{\text{minimize}} f_i(x), \ (i = 1, 2, \ldots\ldots, M) \qquad (7.1.1.1)$$

$$h_j(x) = 0, (j = 1, 2, \ldots\ldots. J) \qquad (7.1.1.2)$$

$$g_k(x) \leq 0 (k = 1, 2, \ldots\ldots. K) \qquad (7.1.1.3)$$

$f_i(x)$, $h_j(x)$, $g_k(x)$ are functions of the design vector $x = (x_1, x_2,........., x_n)^T$.

Each x_i in x is a decision or design variable and can be discrete, continuous, or mixed.

R^n is the search space spanned by the decision variables.

Each $f_i(x)$ is an objective or cost function.

On the basis of objective functions, the problem can be categorized as single or multi-objective optimization problem.

The space spanned by the objective functions is the solution space.

The equalities and inequalities for h_j and g_k are called the constraints.

The above is a generic example of a minimization problem.

If the inequality is changed to $\geq$, the problem becomes a maximization problem.

A solution satisfying all the constraints is a feasible solution. If there are only constraints and no objective, the problem is simply a feasibility problem, where each feasible solution is also an optimal solution. Each of $f_i(x)$, $h_j(x)$, $g_k(x)$ can be linear or non-linear. Most real-world optimization problems are, however, non-linear.

7.1.2 Classification of Optimization Algorithms

A simple way toward classifying optimization algorithms is to split them into two categories: **deterministic** and **stochastic**. Deterministic algorithms are non-random, predictable algorithms by nature, which given an initial input always produce the same output with the same sequence of steps. However, these are found to be insufficient for many real-world problems. Stochastic algorithms, on the other hand, search the solution space with a certain degree of randomness in their searching technique.

Heuristic algorithms discover solutions via the trial-and-error method and provide no guarantee of returning an optimal solution. However, solutions close to optimal solutions can be found in a reasonable amount of time. A major limitation of heuristic algorithms is that they often get trapped at local optima. Metaheuristics, on the other hand, are a step up from the simple heuristics and generate better results. They use randomization to avoid getting trapped at a local optimum.

Metaheuristic algorithms can broadly be sub-divided into single-solution-based and population-based algorithms. Single solution-based algorithms work on a single solution, or agent, where the steps trace a trajectory in the search space; hence, these may also be termed trajectory-based algorithms. In contrast, population-based algorithms work on a set, or population, of given solutions. They can further be classified into evolutionary or swarm-based techniques. Evolutionary algorithms are derived basically from natural processes such as biological evolution, natural selection, mutation, etc. The best solutions are carried forward, having been selected by means of a fitness function. This corresponds to the survival of the fittest mechanism in nature. Swarm-based algorithms make use of collective, decentralized,

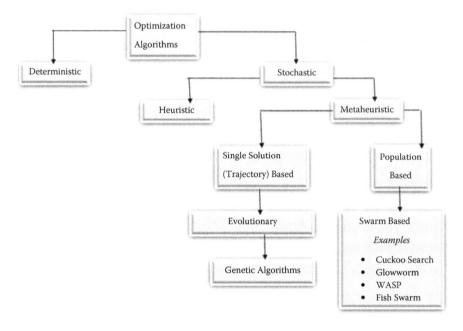

FIGURE 7.1
Classification of optimization algorithms.

distributed and self-organizing agents to arrive at a final solution. Figure 7.1 summarizes the classification of optimization algorithms.

Almost all metaheuristics aim at incorporating two features of natural selection viz. Survival of the fittest and adaption to the environment. These features are, respectively, randomness intensification (or exploitation in a general sense) and diversification (or exploration in a general sense). Intensification is the process searching for the best candidate solutions among current best solutions, and diversification refers to an efficient exploration of the given solution space.

In this chapter we discuss four nature-inspired swarm optimization techniques namely the Cuckoo Search, Glowworm, WASP and Fish Swarm Optimization Techniques. Each algorithm in this chapter is explained conceptually with pseudo-codes, mathematical proofs, and details on the variants.

7.2 Cuckoo Search

7.2.1 Introduction to Natural Cuckoo Behaviour

Cuckoos are a fascinating bird species. Not only can they make diverse sounds or calls, but they also reproduce in diverse ways. Cuckoos construct their own nests, some, like the anis and guira, build communal nests. A substantial minority, including the common cuckoo, indulge in brood parasitism. Brood parasitism is the strategy of raising one's young ones by relying on others. Cuckoos choose the nest of a bird (either of the same or another species), drop their eggs in that nest, and henceforth the cuckoo's egg completely depends on the host's care.

If the host bird identifies that the egg is not one of theirs, either the egg is thrown away or the nest is abandoned and the host-bird re-creates a new nest. To prevent this from happening, the cuckoo females have evolved to make use of egg mimicry. Eggs are distributed across different nests to reduce the chances of egg loss. If the cuckoo's egg is not identified, it generally goes on to hatch before the host bird's eggs and kicks out the other eggs from the nest. This leads to an increase in food share. Some cuckoo chicks can also imitate the call of the host's chicks.

An advantage of brood parasitism is that the parents do not have to invest in building a nest or nourishing their young ones. They can spend more time on hunting and reproducing. Over time, natural selection has made both host birds and the cuckoo evolve so that in each generation the fittest bird survives. This breeding behaviour of some cuckoos is one of the best models for co-evolution and is the basis for a recently developed optimization technique viz, the cuckoo search.

7.2.2 Artificial Cuckoo Search

Brood parasitic cuckoo species breeding behaviour and Lévy flight behaviour of some birds and fruit flies inspired a novel swarm-based search technique, termed **Cuckoo Search** proposed by Xin-She Yang and Suash Deb (2009).

The artificial cuckoo search technique is inspired from the brood parasitic breeding behaviour of cuckoo species and uses the concept of Lévy flights, which are found in the flight behaviour of many animals and insects.

The cuckoo search algorithm (Yang, 2013) is derived from following three rules:

- At a single point of time, every cuckoo lays one egg and dumps the egg in the nest chosen in random manner.
- Only the best nests with high-quality eggs carry forward to the next generations.
- The number of available host nests is fixed, and the egg laid by a cuckoo is discovered by the host bird with a probability $p_a \in [0, 1]$. In this case, the host bird can either throw the egg away or abandon the nest and build a completely new nest.

Before proceeding further into the algorithm, it is important to discuss a few mathematical terms and functions that are regarded as prerequisites to understanding the cuckoo search algorithm.

7.2.2.1 Random Variables

Any random phenomenon gives output as random variables. These values are denoted by X. A random variable is said to be discrete if it takes only distinct values like 1, 2. A random variable is continuous if it can take any of the infinite values over an interval. These are usually represented by the area under a curve or an integral. The probability of a random variable X belonging to a set of outcomes A, is the area above A and under a curve. The total area under this curve must be 1, and it should have no negative value for any element in set A. Such a curve is called a 'density curve.'

7.2.2.2 Random Walks

A random walk (say S_N) is the sum of a series of random steps each represented by a random variable X_i.

$$N = \sum_{i=1}^{N} X_i = X_1 + \ldots\ldots + X_N \qquad (7.2.2.1)$$

Random walks can be of fixed or variable length (depending on the step size). A random walk is used to determine the web size, sample online graphs, or represent node movement in a wireless network.

7.2.2.2.1 Power Law

The power (or scaling) law applies when a relative change in one quantity results in a proportional relative change in another quantity. The general form of a power law distribution is:

$$Y = kX^{\alpha} \qquad (7.2.2.2)$$

Where X and Y are variables of interest; α is the law's exponent; k is a constant.

7.2.2.3 Heaviside Function

It is a mathematical function denoted by H or θ, which can represent either a piecewise constant function or a generalized function. The Heaviside function as piecewise constant function:

$$H(x) = \begin{array}{l} 0 \text{ if } x < 0 \\ 1/2 \text{ if } x = 0 \\ 1 \text{ if } x > 0 \end{array} \qquad (7.2.2.3)$$

The Heaviside function as a generalized function:

$$\int \theta(x)\phi'(x)dx = -\phi(0) \qquad (7.2.2.4)$$

Shorthand notation:

$$H_c(x) \equiv (x-c) \qquad (7.2.2.5)$$

7.2.2.4 Lévy Distribution

The Lévy distribution is a stable cum continuous probability distribution for a non-negative random variable. It is a stable distribution.
   The Lévy distribution can be represented in simple form as follows:

$$L(s,\gamma,\mu) = \begin{cases} \sqrt{\dfrac{\gamma}{2\varPi}}\, exp\left[-\dfrac{\gamma}{2(s-\mu)}\right]\dfrac{1}{(s-\mu)^{3/2}}, & 0<\mu<s<\infty \\ 0 & \text{otherwise} \end{cases} \qquad (7.2.2.6)$$

where, μ is the minimum step and γ is the scale parameter.

7.2.2.4.1 Lévy Flights

Lévy flights are random walks whose step lengths are drawn from a Lévy distribution. Lévy flights are represented by the following formula:

$$L(s) \sim |s|^{-1-\beta} \quad 0<\beta \leq 2, \qquad (7.2.2.7)$$

where β is an index.
   The following demonstrates the pseudo-code for cuckoo search, and Figure 7.2 shows the flow chart representation of the cuckoo search algorithm.

> **PSEUDO-CODE FOR CUCKOO SEARCH ALGORITHM**
> **Begin**
> **Objective Function** f(I), I = (i1, i2, ..., id)$^\text{T}$
> **Generation of Initial Population**
>   for (n) host nests: I$_j$ (j = 1,2,.....,n)
>   **While** (t < MaxGeneration) || (Stop Criteria)
>   **Select** a cuckoo randomly by Lévy flights
>   **Calculate** the quality fitness F$_j$

Choose Randomly a nest among n (suppose, K)
if ($F_j > F_k$),
        replace k by the new solution;
end
A Fraction (p_a) of worse nests
        are thrown off and new ones are built;
Keep the best solutions
        (or nests with quality solutions);
        Solutions are ranked and best solution is considered
    end while
  Postprocess results and visualizations
  end.

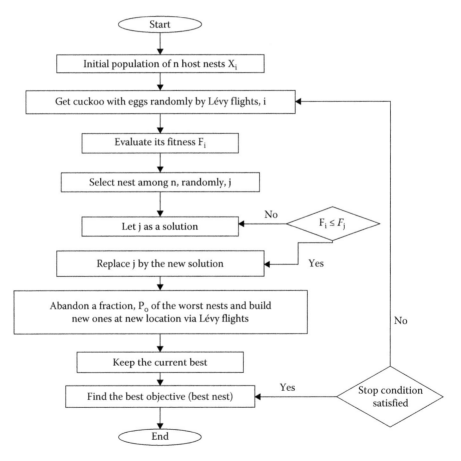

FIGURE 7.2
Cuckoo search algorithm.

7.2.2.5 Advantages of Cuckoo Search

Cuckoo search (CS) has quite a number of advantages, as follows:

- Cuckoo search maintains an effective balance between local search and diversification or randomization for search space exploration.
- Cuckoo search comprises only two control parameters: population size n and probability p_α. This makes the algorithm less complex and more generic.
- If the population size is fixed, it is only p_α that controls the elitism (carrying the best solutions over to the next generation to build new solutions) and randomization balance and local search.
- Cuckoo search has efficient randomization and there is the possibility for large steps.
- Convergence rate is independent of p_α hence, the parameters do not need to be fine-tuned for each new problem.
- It can easily be adapted to varying optimization problems due to its generic nature.
- It is superior to both PSO and GA with regard to multimodal functions.
- It provides best solutions to multi-objective optimization problems.
- It can be easily combined with other Swarm-based techniques like PSO and BCO (Bao & Zeng, 2009; Karaboga & Akay, 2009; Theraulaz, Bonabeau, & Deneubourg, 1998).
- It is highly simple to understand and implement.

7.2.3 Mathematical Proof of Cuckoo Search

As mentioned before, the cuckoo search performs a global search or exploration via Lévy flights with a variable step size with intermediate 90-degree turns. Occasional large steps ensure that the cuckoo search algorithm doesn't get trapped in a local optimum.

The algorithm performs two types of searches: local and global.

Local search is performed via local random walks as:

$$x_i^{(t+1)} = x_i^{(t)} + \alpha s \otimes H(p_\alpha - \varepsilon) \otimes \left(x_j^{(t)} - x_k^{(t)} \right) \qquad (7.2.3.1)$$

x_j and x_k are solutions selected randomly;

H(u): Heaviside Function;

p_α: Switching parameter for balancing local and global random walks;

s: Step size;

ε : is a random number from uniform distribution.
   Global search is performed using Lévy flights as follows:

$$x_i^{(t+1)} = x_i^{(t)} + \alpha\, L(s,\lambda) \qquad\qquad (7.2.3.2)$$

where $\alpha > 0$ is the step size scaling factor and

$$L(s,\lambda) = \frac{\lambda\Gamma(\lambda)\sin(\pi\lambda/2)}{\pi}\,\frac{1}{s^{1+\lambda}} \quad (s \gg s_o > 0; s_o \text{ being the smallest step})$$
$$(7.2.3.3)$$

The Lévy flight is a random walk with step length drawn from a Lévy distribution.

$$\text{Lévy} \sim \frac{1}{s^{\lambda+1}} \quad 0 \le \lambda \le 2 \qquad\qquad (7.2.3.4)$$

Random numbers generated from Lévy flights consist of a random direction choice, which is created via uniform distribution, and steps are generated using Mantegna's algorithm.
   In Mantegna's algorithm,

$$s = \frac{1}{|v|^{1/\beta}} \qquad\qquad (7.2.3.5)$$

where u and v are drawn from a normal distribution such that,

$$u \sim N(0,\ \sigma_u^{\,2})\ and\ v \sim N(0,\ \sigma_u^{\,2}) \qquad\qquad (7.2.3.6)$$

$$\sigma_u = \left\{ \frac{\Gamma(1+\beta)\sin\left(\frac{\beta}{2}\right)}{\Gamma\left[\frac{1+\beta}{2}\right]\beta\,2^{(\beta+1)/2}} \right\}^{1/\beta} \quad and\ \ \sigma_v = 1 \qquad (7.2.3.7)$$

Below is a brief mathematical analysis of the global search.
   Let $r\ \hat{e}[0,1]$ be a random number for the global branch and p_α be the switching or discovering probability such that,

$$\begin{aligned} x_i^{(t+1)} &\leftarrow x_i^{(t)} & if\ r < p_\alpha \\ x_i^{(t+1)} &\leftarrow x_i^{(t)} + \alpha \otimes L(s,\lambda) & if\ r > p_\alpha \end{aligned} \qquad (7.2.3.8)$$

Steps for the cuckoo search algorithm (Yang, 2013) are:

1. Initial population of n nests is generated at positions $X = \{x_1{}^0, x_2{}^0,$ ...., $x_n{}^0\}$ randomly; after that, objective values are evaluated to determine the current global best $g_t{}^0$.

2. New solutions are updated as:

$$x_i^{(t+1)} = x_i^{(t)} + \alpha \otimes L\,(\lambda) \qquad\qquad (7.2.3.9)$$

3. A random number r is drawn from a uniform distribution [0, 1]. Update $x_i^{(t+1)}$ if $r > p_\alpha$. In order to determine the new global best $g_t{}^*$, new solutions are evaluated.
4. Once the stopping criterion is reached, $g_t{}^*$ can be claimed as the best global solution found so far. Otherwise, return to step 2.

7.2.4 Variants of Cuckoo Search

7.2.4.1 Discrete Cuckoo Search

Discrete cuckoo search aims to solve combinatorial problems like the travelling salesperson problem (TSP), scheduling problems, etc. Combinatorial problems are those where the number of combinations increases exponentially as the problem size grows. Solving these problems results in unrealistic and expensive computations because there might exist some extreme combinations.

Ouaarab, Ahiod, and Yang (2014) proposed an improvised and discrete version of the cuckoo search for the TSP. Ouyang et al. (2013) have proposed a discrete version for the spherical TSP. TSP can be defined as an NP-hard combinatorial optimization problem, where a salesman visits a finite number of cities, such that each city is covered only once, and then finally returns to the initial city. The tour of the salesman should be constructed such that the overall cost and distance travelled are minimized. The authors proved that their discrete version of the CS algorithm can be adapted to other optimization problems by reconstructing the population.

7.2.4.2 The Improved Cuckoo Search

The proposed improvement considers a cuckoo as a first level for controlling intensification and diversification, and the population is considered to be a second controlling level. A new type of smarter cuckoo is added for population restructuring. These new cuckoos are capable of changing their host nests to reduce the chances of abandonment. We thus have a new fraction of cuckoos p_c, which first search for a host nest (new solution) via Lévy flights and then from the current solution seek a better solution.

The following demonstrates pseudo-code for the improved cuckoo search.

Begin
Objective Function $f(I)$, $I = (i1, i2, \ldots\ldots, id)^T$
Generation of Initial Population
      for (n) host nests: I_j ($j = 1,2,\ldots,n$)
    While ($t < $ MaxGeneration) || (Stop Criteria)
        Search with a fraction(Pc) of smart cuckoos
        Choose a cuckoo randomly using Lévy's Flight
          Calculate the quality/ fitness F_j
            Select a nest from hn (e.g. K) randomly
      if ($F_j > F_k$),
        replace k by the new solution;
      end if
      A Fraction (p_α) of worse nests
        is thrown off and new ones are built;
      Keep the best solutions
        (or nests with high-quality solutions);
      Solutions are ranked and best solution is considered
    end while
Postprocess results and visualizations

The authors have used perturbations in the algorithm. The use of local perturbations makes the algorithm flexible enough to be adapted for other optimization problems.

7.2.4.3 Binary Cuckoo Search

Optimization problems can be either continuous in which the solutions are represented by a set of real numbers or discrete and can be represented by a set of integers. In binary optimization, however, the solutions are represented by the set of bits. The binary version of the cuckoo search (BCS) can be applied to problems in routing, scheduling, feature selection, etc.

In the original CS algorithm using Lévy flights, the solutions are represented as a collection of real numbers present at a continuous search space. These are required to be converted into binary values that can adapt to the discrete binary version of the cuckoo search. The BCS has two main aspects to it (Pereira et al., 2014):

- Main binary cuckoo dynamics, which include:

  - Obtaining new solutions using Lévy flights.
  - Binary solution representation (BSR) to find possibilities of flipping of each cuckoo.

- Objective function and selection operator – the principles are same as those of genetic algorithms:

Let x_i be a continuous valued solution in the interval [0, 1] and x_i' be a binary solution representation. Using the sigmoid function to convert values we have:

$$S(x_i) = \frac{1}{1 + e^{-x_i}} \qquad (7.2.4.3.1)$$

where $S(x_i)$ represents the flipping chance of x_i'.

To determine binary solution xi', $S(xi)$ is matched with the result of a generated random number γ obtained from the interval [0, 1] for each dimension i belonging to the solution x. If the flipping chance is greater than γ, the binary value is assigned is 1, else 0:

$$x_i' = \leftarrow \begin{cases} 1 & \text{if } \gamma < S(x_i) \\ 0 & \text{otherwise} \end{cases} \qquad (7.2.4.3.2)$$

Pereira et al. (2014) proposed a BSR algorithm for feature selection. Pseudo-code is demonstrated as follows.

Pseudo-Code for Binary Cuckoo Search for Feature Selection
Input: Labeled Training set A_1 and Evaluating set A_2, loss parameter L_1 a value, number of nest N, dimension d, number of iterations T, c1 and c2 values.
Output: Global best position

```
for each nest N1 (∀i = 1,........,n)
      for each dimension D (∀D = 1........d)
      k¹ (0) ← Random {0,1};
   end
   f_c ← -∞;
end
global fit ← -∞ ;
for every iteration (i= 1,..........,I)
      for all nest Ni (∀i = 1.......n)
            Create A₁ and A₂ from A₁ and A₂ such that the set carries
            features
            For N1 where k1 ≠ 0 ∀j = 1........d;
            Training of OPF over A1 and evaluation over A2 and
            storing the
               accuracy over ACC;
               if ACC > fl← ACC;
                  for every dimension D(∀D = 1........d)
                  xl^←xl(t);
            end
```

```
        end
      end
      [maxfit, maxindex] ← max(f);
      if(maxfit>globalfit)
         globalfit√ maxfit;
         for all dimension j (∀j = 1.......d)
         ĝi ←xl maxindex(t);
         end
      end
      for all nest Ni(∀i = 1.......n)
          for each dimension D(∀D = 1.......d)
             identify worse nest having pa⊕ Lévy(Υ);
```

$$\text{if}\left(\sigma < \frac{1}{1+e^{xl\,c^t}}\right)\text{then}$$

```
             xl(t) ← 1;
             else xl(t) ← 0;
                 end
             end
         end
  end
```

A few of the researchers' contributions using the proposed algorithm are: Soto et al. (2015) for solving the set covering problem, Gherboudj, Layeb, and Salim (2012) for solving the 0–1 knapsack problem. Feng, Ruan, and Limin (2013), who have also proposed a binary cuckoo search algorithm.

7.2.4.4 Chaotic Cuckoo Search

In the chaotic cuckoo search (CCS), the characteristics of chaos theory are incorporated into the cuckoo search technique. Chaos theory studies the behaviour of highly sensitive systems in which slight changes in the initial position can make big differences to the system's behaviour. Chaos has the properties of non-repetition and ergodicity, which facilitate faster searches. The concept of elitism from genetic algorithms is also incorporated to carry the best (fittest) cuckoos over to the next generation to build new and better solutions. The CCS algorithm proposed by Wang et al. (2016) uses a chaotic varying step size α, which can be tuned by chaotic maps (exhibiting chaotic behaviour) after normalization. By normalizing all chaotic maps, their variations are always in the interval [0,2]. Elitism prevents the best cuckoos from corruption as the solutions are updated.

The following demonstrates the pseudo-code for chaotic cuckoo search

Pseudo-code for Chaotic Cuckoo Search
Begin:
    Step 1: Initialization. Set Generation counter i = 1;
    Randomly Initialize the population P

Randomly Set p_a, initial value of the chaotic map c_0, and elitism parameter
   KEEP.
Step 2: **While** t < MaxGeneration
   the population should be sorted according to their fitness
   KEEP ← Best Cuckoos
   Update step size using chaotic maps ($\alpha = c_i + _1$)
   Select a cuckoo randomly (suppose, i) and replace the solution by
   performing
   Lévy flights
   Calculate its fitness F_i
   Choose randomly a nest among n (suppose, k)
   If ($F_i < F_k$)
       Replace k by new solution.
   End if
   A part (pα) of the worse nest is replaced with new ones built
   Change the KEEP worse cuckoo with the KEEP best cuckoo.
   Sort the population to determine the current best.
       t = t+1
Step 3: end while
End

7.2.4.5 Parallel Cuckoo Search

Parallelization is a natural extension for nature-inspired population-based algorithms. Tzy-Luen et al. (2016) proposed the parallel cuckoo search. This algorithm divides the main population into subpopulations, thus increasing the diversity in exploring the solution space. The algorithm outputs the fittest host x_i.

Pseudo-Code for Parallel Cuckoo Search
Begin
Objective Function f(n), n= $(n_1, \ldots \ldots, n_d)_T$
Generate initial main population for n number of host nest np (i=
   1,......,p)
Population is divided into subpopulation where subpopulation 1.....p executes the cuckoo-
   search on the same search space parallel
while (i < stop criteria)
   Subpopulation 1......p in parallel
   Select a cuckoo randomly by Lévy Flight F_i
   Calculate its quality / fitness F_i
   Choose a nest among p (suppose, s) randomly
   If ($F_i > F_v$) then

> replace s by the new solution
> **end**
>   A fraction (P_a) of worse nests are abandoned and new nests are
> built
>   Preserve the best solution (nests)
>   Rank solutions and find the best available
>   Synchronize fittest nests at every n generation
> **End**
> **End**

7.2.4.6 Gaussian Cuckoo Search

The basic cuckoo search algorithm efficiently finds the optimum solution, but it does not guarantee fast convergence and precision. With this aim in mind, Zheng and Zhou (2012) introduced some modifications to the original algorithm, which are:

$$\sigma_s = \sigma_0 \exp(-\mu k) \qquad (7.2.4.6.1)$$

where σ_0 and μ are constants, and k refers to the current generation.
The new solution is thus generated as:

$$x_i^{(t+1)} = x_i^{(t)} + \alpha \otimes \sigma_s \qquad (7.2.4.6.2)$$

with α being the step size related to the scales of the problem. Mostly it is taken to be 1. The pseudo code for Cuckoo search based on Gaussian distribution is as follows.

> **Pseudo-code for Gaussian Cuckoo Search**
> **Begin**
>   **Objective Function** f(n), n= (n1,. . . . ., nd)T
>   **Generation** of initial population for h number of host nest ni
>   (1,. . . . .,h)
>   **While (t<MaxGeneration) or (Stop Criterion)**
>   **Select** a Cuckoo randomly by Gauss distribution
>   **Calculate** the Quality/fitness function fi of the cuckoo
>   **Randomly** Choose a nest among h (suppose, k)
>   **If** (Fi >Fk)
>   **Replace** k with the new solution
>   **End**
>   A portion of worse nests (Pα) are abandoned and replaced by
>   newly built ones;
>   Preserve the best solution (or high-quality solutions)
>   Rank the solutions according to quality and find the current best;
>   **End While**
>   Postprocess results and visualizations
> **End**

7.2.5 Application of Cuckoo Search in Real-World Applications

Cuckoo search finds its applications in varying domains of optimization, computational intelligence, etc. as elaborated below:

1. *Medical Applications:*

   - A hybrid model of the cuckoo search algorithm (CSA), support vector machines (SVM), and PSO was developed by Liu and Fu (2014), which achieved better accuracy than the PSO-SVM and GA-SVM models for classification when applied to heart disease and breast cancer datasets.
   - Giveki, Salimi, Bahmanyar, and Khademian (2012) used a modified CSA in combination with feature-weighted SVMs to develop a novel approach for early diabetes diagnosis, which achieved 93.58% accuracy when applied to the CI dataset.

2. *Clustering:*
   Clustering is the segregation of patterns/objects into groups, or clusters, based on their similarities and differences.

   - A hybrid of PSO and CS techniques provides an increased cover of the c-means defects along with reduced compilation time, fewer iterations, and faster convergence.
   - CSA and its hybridizations have been shown to be successful in clustering of web search engine results by Cobos et al. (2014) and web documents by Zaw and Mon (2013).
   - In another work, Goel, Sharma, and Bedi (2011), proposed an improvised biomimicry technique by combining CSA with an unsupervised classification method, which was applied to two image datasets of real-time remote monitoring satellites for extraction of water body data.

3. *Image Processing:*
   Image processing is used to extract useful information from images. A popular application is face recognition.

   - Bhandari et al. (2014a) combined CSA with wind-driven optimization along with Kapur's entropy for efficient use in multilevel thresholding, thereby improving information extraction from satellite images.
   - In another work by Bhandari et al. (2014b), the authors integrated discrete wavelet transform [DWT] with CSA to improvise satellite images of low contrast.
   - An image multi-thresholding based on CSA was used by Raja and Vishnupriya (2016) to find an optimized threshold value for

segmenting an RGB image. This algorithm fared better than the previously applied firefly and PSO algorithms.

4. *Engineering design:*
   It is the most popular application domain.

   - Yang and Deb (2013) proposed a multi-objective CSA for optimization of structural design problems like disc brake and beam design.
   - Meziane, Boufala, Amar, and Amara (2015) proposed a novel methodology using CSA for the wind power system design optimization.
   - Also, Yildiz (2013) applied CSA to the manufacturing optimization for the problem of milling.

5. *Power and Energy systems:*
   In this domain, CSA has been applied to:

   - Optimal capacitor allocation
   - Power transmission network
   - Resource operation optimization
   - Location of maximum power point in PV solar energy systems.

6. *Scheduling:*

   - Nurse scheduling
   - Flow shop scheduling
   - Short-term fixed-head hydrothermal scheduling
   - Short-term cascaded hydrothermal scheduling

7. *Software:*

   - Srivastava, Chis, Deb, and Yang (2012) used the cuckoo search for the generation of independent paths and test data in software testing.

7.3 Glowworm Algorithm

7.3.1 Introduction to Glowworm

Glowworms refer to the group of insects that produce and emit light, which are also known as lightning bugs or fireflies. The best time to catch sight of glowworms' sparkling magical lights is a dark summer night, as can be seen from Figure 7.3.

FIGURE 7.3
Glowworm.

They use a process called bioluminescence to emit light. However, many living creatures have been discovered who exhibit similar sparkling behaviour such as jellyfish, some kinds of bacteria, protozoa, hydrozoa etc. In fact, approximately 80 to 90 percent of ocean life comprises bioluminescent creatures. The reason behind the popularity of glowworms is that they are found easily and in large numbers.

Catching a glimpse of a glowworm winking on and off on a summer night is easy compared to getting to see their swarming behaviour. A large number of glowworms congregate to form a swarm and emit flashes of light synchronously, as shown in Figure 7.4. After having witnessed these beautiful sights, one wonders about the "hows" and "whys" of this light-emitting and swarming behaviour of glowworms. The reasons for this light-emitting behaviour are to attract their unaware prey to their trap and to attract a mate for reproduction. In the process of reproduction, individual courtship and mass mating are both observed. How photic communication and other mechanisms take place in glowworm swarms is discussed next because our aim is to develop artificial swarm systems based on their behaviour.

The life cycle of a glowworm begins with an egg, then it transforms from egg to pupa, pupa to larva and larva to adult. A glowworm gets only a few weeks to be an adult. Therefore, it faces an urgent need to discover a mating partner to reproduce with an objective of species preservation.

FIGURE 7.4
Glowworm swarm.

7.3.1.1 Flashing Patterns

With evolution, glowworms have evolved in a way that they can control the emission of light in a number of ways so as to generate different mating signals. They generate different signals by varying the parameters mentioned such as:

- Colour of light emitted
- Brightness of light emitted
- Male flashes and female flashes – phase difference
- Duration of each flash
- Number of flashes per cycle
- Time period of flashes
- Continuous glow or pulse train of flashes

Kaipa and Ghose (2017) describe these flashing patterns with various examples such as in *Lampyrusnoctiluca*, a species of glowworm found commonly in Europe, in which only the apterous female has the ability to glow. She wriggles over grass and sweeps her light from one direction to another to draw the attention of wandering male glowworms. In the *Lamprophorusteneb-rosus* glowworm, light emitting capability is present in both males and females. The female does not have wings and it uses its light in a very similar way to *Lampyrus*, to attract a mate. In some species, the female uses different patterns such as glowing with long flashes, incompletely extinguished during the intervals. When the males sense such a pattern from as far as 10 feet away, they fly down to the female. These flashing patterns are species specific. For example, in the glowworms that belong to the genus *Photinus*, the male initiates mating by crawling over the ground, emitting light in various patterns, and observing the response signals of the females nearby. In *Photinus consanguineus*, two flashes are emitted by the male and then the male takes a pause, followed by two more flashes. It repeats this pattern. The female replies within a second after the second flash of the male. In *Photinus castus*, the male emits a long single flash and the female replies immediately to it. Although the structures of *P. consanguineus and P. castus* are very similar, they are considered different species on the basis of their different patterns of emitting light – but very often they are found flying together. However, not a single case of interbreeding has been observed. This is an extremely remarkable feature of glowworms, which they achieve by having the ability to emit light in various patterns and recognize them.

7.3.1.2 Mass Mating

The species mentioned above have one thing in common, either the male or the female, whoever is attracting the other one, needs an uninterrupted line of vision so that they can perceive signals and reply. In some visually cluttered areas, such as the mangrove swamps of southeast Asia, getting such an uninterrupted line of sights is not easy. Therefore, the species found in such regions do not have the option of individual courtships. So, the species of glowworms here form swarms on trees or caves because then it is easy for wandering glowworms to find a mating partner. If initially there is no swarm, then through mutual photic attraction glowworms might form a nucleus. There is competition while forming these swarms, which leads to more than one larger swarm that further draws nearby smaller nuclei because of higher mean light emission, which indicates a higher probability of finding a mate.

7.3.2 Glowworm Swarm Optimization (GSO) Algorithm

The glowworm Swarm Optimization (GSO) algorithm is developed on the basis of behaviour observed in glowworms and glowworm swarms. GSO was proposed by Krishnanand and Ghose, (2005), who used this algorithm in numerous applications. Initially, the goal of developing GSO was to propose solutions to numerical optimization problems instead of identifying the global optimum. But GSO has made a lot more contribution in areas such as in robotics due to its decentralized decision making and movement protocols.

Initially, GSO randomly distributes a group or swarm of agents in random search space, inspired by glowworms. These agents are embued with other behavioural mechanisms that are lacking in their natural counterparts. The basic working of the algorithm is based on three mechanisms mentioned below:

1. Fitness broadcast
Glowworms have a pigment known as *luciferin*, which enables a glow-worm to illuminate. The amount of luciferin a glowworm has defines the fitness of their locations in the objective space.

2. Positive taxis
A glowworm is attracted to a neighbour that is glowing brighter than itself, and therefore starts moving toward it.
   When there are more than one such neighbour it makes use of a probabilistic mechanism to select one.

3. Adaptive neighbourhood
Each glowworm make use of an adaptive neighbourhood for identifying neighbours, which is defined by a local-decision domain having a variable range r_d^i bounded by a hard-limited sensor range rs ($0 < r_d^i \leq r_s$). A suitable heuristic can be used here to modulate r_d^i. The glowworms solely depend upon the local information for movements. Each glowworm selects a neighbour carrying a luciferin value greater than its own, thereafter moving toward it. These movements are built upon available local infor-mation as well as selective neighbour interactions that enable the swarm of glowworms to cluster into disjoint subgroups drifting toward, and meeting at, multiple optima of a given multimodal function.

7.3.2.1 Algorithm

Although the algorithm is explained for finding multiple optima of multi-modal functions, it is very easy to modify it for minimization problems. Initially, GSO places glowworms randomly in the search space such that they are well dispersed. Initially, the amount of luciferin in each glow-worm is equal to zero. A unit cycle belonging to the algorithm consists of a luciferin update phase, a movement phase, and a neighbourhood range update phase, as shown in Figure 7.5.
   The following demonstrates GSO algorithm:

> **Glow Swarm Optimization**
> **Begin**
> **Set** Number of Dimensions = d
> **Set** Number of Glowworms = g
> Let step size = n
> Let $x_i(t)$ is location of Glowworm i at time t
> **Deploy** the agents randomly;
> **For** i= 1 to g do $L_i(0)$ = L0

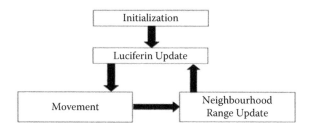

FIGURE 7.5
GSO Algorithm phases.

$r_d = r_0$
Set the Maximum Iteration = iter max;
Set t = 1;
While (t ≤ iter max) do;
{
    for each glowworm i
        i(t) = (1 - ρ) i (t -1) + yJ(xi (t));
        for each glowworm i
        {   Ni(t) = {j : dij(t) <rid(t) ;
        i(t) j(t)
        }
    for each glowworm j ε Ni(t)
        pij(t) = j(t) − i(t)
        k ε Ni(t) k(t) − i(t)
        j = select glowworm ();
        xi(t+1) = xi(t) + n xj (t) − xi(t)
        xj(t) − xi(t)
        rid
        (t+1) = min { rs, max { 0, rid (t) + β (nt − |Ni(t)|)}};
    }
    t ← t +1;
end

7.3.3 Mathematical Proof of GSO Algorithm

Krishnanand and Ghose, (2005) derived certain theoretical proofs about the working procedure of luciferin update rule. The results of these proofs show the variations in quantity of luciferin defined as a function of time, while the algorithm executes.

We discuss these theoretical proofs next in detail as Theorems 7.3.3.1 and 7.3.3.2.

The two things proved here are:

- Due to luciferin decay, the maximum luciferin level τ_{max} is bounded asymptotically.
- The luciferin l_j of all glowworms co-located at a peak Xi converge to the same value.

Theorem 7.3.3.1 Assuming that the luciferin update rule is used, the luciferin level $l_i(t)$ for any glowworm i is bounded above asymptotically as follows:

$$\lim_{t \to \infty} l_i(t) \le \lim_{t \to \infty} l^{max}(t) = \left(\frac{\gamma}{\rho}\right) J_{max} \tag{7.3.3.1}$$

where, J_{max} is the global maximum value of the objective function.

Proof
Given, the maximum value of objective function as J_{max} along with the luciferin update rule in (7.8.1), the max quantity of luciferin added to the existing luciferin level at any iteration t is γJ_{max}. Hence, at iteration 1, the maximum luciferin of any Glowworm i can be calculated as $(1 - \rho) l_0 + \gamma J_{max}$ at iteration 2, $(1 - \rho)^2 l_0 + [1 + (1 - \rho)]\gamma J_{max}$, and so on. The generalized formula, at any iteration t, for maximum luciferin $l^{max}(t)$ of any Glowworm can be given as:

$$l^{max}(t) = (1 - \rho)^t l_0 + \sum_{k=0}^{t-1} (1 - \rho)^k \gamma J_{max} \tag{7.3.3.2}$$

Clearly,
$$l_i(t) \le l^{max}(t) \tag{7.3.3.3}$$

Since $0 < \rho < 1$, from above equation we can calculate:

$$as \ t \to \infty, \quad l^{max}(t) \to \left(\frac{\gamma}{\rho}\right) J_{max} \tag{7.3.3.4}$$

Theorem 7.3.3.2 "For all glowworms i, co-located at peak-locations X_j^* associated with objective function values $J_j^* \le J_{max}$ (where $j = 1, 2, \ldots, m$, with m as the number of peaks), if the luciferin update rule in (7.8.1) is used, then $l_i(t)$ increases or decreases monotonically, and asymptotically converges to $l_j^* = \left(\frac{\gamma}{\rho}\right) J_j^*$."

Proof

According to (7.8.1), $l_i(t) \geq 0$ always. The stationary luciferin l_j^* associated with peak j satisfies the following condition:

$$l_j^* = (1 - \rho)l_j^* + \gamma J_j^* = \left(\frac{\gamma}{\rho}\right)J_j^* \qquad (7.3.3.5)$$

If $l_i(t)$ l_j^* for Glowworm i co-located at peak-location X_j^*, then using the above equation we have

$$J_j^*(t) > \left(\frac{\rho}{\gamma}\right)l_i(t) \qquad (7.3.3.6)$$

Now,

$$l_i(t+1) = (1 - \rho)l_i(t) + \gamma J_j^*$$

$$> (1 - \rho)l_i(t) + \gamma\left(\frac{\rho}{\gamma}\right)l_i(t) \qquad (7.3.3.7)$$

$$\Rightarrow \quad l_i(t+1) > l_i(t) \qquad (7.3.3.8)$$

that is, $l_i(t)$ increases monotonically.

If $l_i(t) > l_j^*$ for Glowworm i co-located at peak-location X_j^*, then using the above two equations it is easy to show that

$$l_i(t+1) < l_i(t) \qquad (7.3.3.9)$$

that is $l_i(t)$ decreases monotonically.

The convergence of sequence $l_i(t)$ is hence proved by showing the fixed point l_j^* of the system. From eq. 7.3.3.9 above, the following relation can be deduced:

$$\left|l_i(t) - \frac{\gamma}{\rho}J_j^*\right| = (1 - \rho)\left|l_i(t-1) - \frac{\gamma}{\rho}J_j^*\right| \qquad (7.3.3.10)$$

$$= (1 - \rho)^2\left|l_i(t-2) - \frac{\gamma}{\rho}J_j^*\right|$$

Proceeding in a similar way, we get

$$|l_i(t) - l_j^*| = (1 - \rho)^t|l_i(0) - l_j^*| \qquad (7.3.3.11)$$

Therefore,

$$lim_{t\to\infty} |l_i(t) - l_j^*| \;=\; lim_{t\to\infty} (1 - \rho)^t |l_i(0) - l_j^*|$$
$$= \; 0, \text{ since } 0 < (1 - \rho) < 1$$

(7.3.3.12)

From the above equation, it is clear that the luciferin $l_i(t)$ of Glowworm i, co-located at a peak-location X_j^* asymptotically converges to l_j^*.

7.4 Wasp Swarm Optimization

7.4.1 Introduction to Wasp Swarm

Wasp colony optimization is another SI technique (Bonabeau, Dorigo, & Theraulaz, 1999; Iourinski et al., 2002) inspired by wasp species and how they collectively work to complete the life-cycle stages such as foraging, brooding, or making nests. Wasps differ from bees in some of their characteristics, and therefore the optimization technique using wasps is not defined under bee optimization, although wasp optimization is described as a subpart of the bee optimization algorithm.

7.4.2 Wasp Colony Optimization (WCO)

WCO provides a strong base for solving various problems with regard to discrete optimization, task assignment, scheduling, etc.

For insects in eusocial societies as discussed by Theraulaz, Goss, Gervet, and Deneubourg (1991), Beckers, Deneubourg, Goss, and Pasteels (1990), all the individuals must cooperate to perform a certain number of tasks, the nature of which depends on the internal needs of the colony, as well as on the particular environmental conditions. At any time, each individual can act and interact either with other individuals or with its environment and thus causes changes in the state of the group.

All social insects, such as termites, ants, and wasps live in organized communities where every member is dependent on the others for survival.

The wasp colony comprises queens (fertile females), workers (sterile female), and males. In late summer, the queens and males mate; the males and workers die off when the weather becomes cold, and the fertilized queen hibernates through the winter. In the spring, plant fibers and other cellulose material are collected by the queen and mixed with saliva to develop a paper-type nest. Figure 7.6 shows wasp colony at different stages.

Wasps take very good care of their nest, although it is built for only one season and houses 10,000 to 30,000 individual wasps. Wasps are highly beneficial because they feed on other insects. A species known as *Polistesdomi-nulus*, Beckers et al. (1990) creates a new colony, and a nest is created of wood

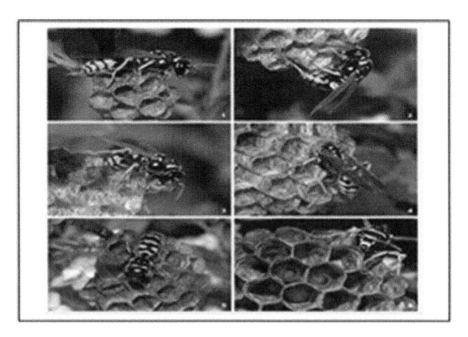

FIGURE 7.6
Stages of forming a wasp colony.

fibers and saliva. Eggs are laid and fertilized. A few weeks after the creation of the nest, new females will emerge, and the colony is expanded.

Theraulaz et al. (1991) introduced the concept of the "organizational characteristic" in a wasp colony. In addition to performing varied tasks like foraging and brooding, wasp colonies organize themselves in definite hierarchy via interactions between individuals, Theraulaz et al. (1998), Litte (1977). The hierarchy starts with the most dominant wasp and moves down to the least dominant wasp, on the basis of their importance and role in the entire colony. For example, if a wasp colony has to fight a war against an enemy colony, wasp soldiers will automatically get more food compared to others because their role is highly important to protect the nest as well as species.

7.4.3 Wasp Algorithm

Wasp algorithms solve different optimization problems with regard to distributed factory coordination, as discussed by Cicirello and Smith (2001), Cicirello, and Smith (2002) Cicirello and Smith (2004)). Job-shop dynamic scheduling was explained by Cao, Yang, and Wang, (2008), self-organization of robot groups by Thereaulaz et al. (1991), distribution of tasks in a multiagent e-learning system by Simian, Simian, Moisi, and Pah (2007), distribution of patients in a health sanatorium system by Cicirello and Smith (2004), dynamic allocation of tasks and resources in a monitoring process within a site, Natura 2000, by Simian, Stoica, and Curtean-Baduc (2008), etc.

Wasp colony optimization (WCO) works as follows:

During the evolutionary process of every generation, a few best fighter wasps are detected on the basis of a high fitness value. Then, the neighbour of every fighter wasp is searched out within a defined radius on the basis of a user-defined parameter. The rest of the wasps die, like male wasps in a wasp colony, because they are no longer required for reproduction of the next generation. As a process of creating the next generation, every fighter wasp may take a lead role along with its respective closest neighbour, inheriting the features that have a high preference value associated with them.

WCO is initialized with random wasps, and every wasp undergoes constraints checking. WCO then determines the optimal wasp by updating the generations. The process of updating occurs as follows:

- All the fighter wasps update among other fighters to generate next generation.
- All non-fighter wasps can update fighter wasps, if any, in their neighbouring community to contribute toward the next generation.
- To compensate for dead wasps, new wasps are created via crossover among fighters.
- The process continues until it converges to an optimal solution or a maximum number of iterations are done.

The general algorithm for wasp swarm optimization can be explained as follows:

Begin
Initialize the population P of wasp with random values;
Number of Iteration is I
For (i= 1 to I)
    Calculate the fitness or function value for each wasp
    Find out the fighter wasp
    Find out its nearest neighbours
    Replace the dead wasp with the newly generated wasp
End
End

7.4.4 Mathematical Proof of the Algorithm

We can get an idea from the algorithm of the scheduling problem that was explained before that the entire updating and learning process depends on the general formula of swarm intelligence, which has been used in the wasp swarm optimization. Therefore, instead of deriving a mathematical proof of the whole algorithm, we can check whether the formula that we have used is optimal or not.

Next, we will see the proof for the optimality of the general formula discussed $(P(r) = \frac{r^2}{1+r^2})$

Theorem: Every optimal function $P(r)$ has the form

$$P(r) = \frac{r^\alpha}{c + r^\alpha} \tag{7.4.4.1}$$

where c and α are any numbers.

$$P = \frac{R_t(T)^\alpha}{R_t(T)^\alpha + c.\theta_t(A)^\alpha} \tag{7.4.4.2}$$

For the optimal function $P(r)$, we have

$$P = \frac{R_t(T)^\alpha}{R_t(T)^\alpha + .\theta_t(A)^\alpha} \tag{7.4.4.3}$$

Considering $\theta' = c^{\frac{1}{\alpha}}.\theta$, the above formula simplifies into

$$P = \frac{R_t(T)^\alpha}{R_t(T)^\alpha + .\theta_t(A)^\alpha}$$

The above formula is minor generalization of the original formula $(P(r) = \frac{R_t(T)^2}{R_t(T)^2 + \theta_t(A)^2})$

which we had considered earlier, and is optimal. We can say this because we have a definition of an optimal function, which is given below.

Definition: A monotonic function $P(r):[0,1]$ is called optimal if, for every $\lambda > 0$, there exists values $a(\lambda)$ and $b(\lambda)$ for which

$$P(\lambda.r) = \frac{a(\lambda).P(r)}{a(\lambda).P(r) + b(\lambda).\left(1 - P(r)\right)}$$

The idea that a probability P should depend on ratio r ($r = \frac{R_t(T)}{\theta_t(A)}$, where $R_t(T)$ is the degree of relevance, and $\theta_t(A)$ is the threshold) is very convincing which has been justified by researchers like Iourinski et al. (2002).

The factors used in the algorithm, such as the ratio r, constant c, threshold, etc., can be modified depending on the type of problem and degree of accuracy. The proof and formulation may slightly differ from problem to problem, but the general approach remains the same and thus the formula we have just proved is optimal and will be used in every case.

7.4.5 Hybrid Wasps

Hybrid approaches can be used sometimes for solving any particular task or problem. Hybrid wasps can be defined as a technique where combination of wasp optimization and any other method, such as learning automata or some other optimization like bee or ant, is used. Practically, there are certain types of parameters on which a particular problem depends on, and it may happen that combining two techniques can give a much better result in terms of accuracy and efficiency. Therefore, researchers are trying to develop such hybrid methods continuously; Simian et al. (2010) in their paper discuss work that has been going on in this field. Relating to biology, hybrid wasps are the types of wasps that are the product (result) when two different species of wasps mate. So, basically, those wasps will have some properties of both the parent (species). This idea is taken in artificial intelligence, or more specifically in swarm intelligence, to combine the different behaviour models of social insects and animals like ants, fish, and wasps to make a new kind of model, which will reflect characteristics of all those.

Various algorithms have been developed based on certain behaviours of wasps such as the recommendation system, multi-agent system, discussed by Dehuri, Cho, & Ghosh (2008); logistic systems explained by Pinto, Runkler, and Sousa (2005); optimization of kernel based on a hybrid wasp clustering model; which is given by T. A. Runkler (2008), etc. Wasp optimization can be applied in data mining and text mining too for better classification and for feature extraction.

In most places, WSO works very well and gives better results compared to the solutions provided by local search (LS) algorithms. There are basically a few types of behaviour according to which we categorize the problem as constructive, navigational, or clustering. So, the problem solving approach should be to first analyze the type of the problem properly and know what behaviour it exhibits and then select the suitable optimization algorithm. There is much more to know about these algorithms, and a lot of work is still required to be done.

7.4.5.1 Application of Wasp Swarm Optimization

There are various fields and problems where wasp swarm optimization has an application. Some of the applications are:

- Scheduling problems
- Task assignment problems
- Robotics

Apart from these applications, it also plays an important role in the medical field, e.g., in health sanatorium systems and in multi-agent systems.

7.5 Fish Swarm Optimization

7.5.1 Introduction to Fish Swarm

The artificial fish swarm algorithm (AFSA) was proposed by Li Xiao-lei in 2002 (Yazdani, Toosi, & Meybodi, 2010), with the aim of imitating fish behaviours like preying, swarming, following, moving, etc. AFSA is based on the collective movement of fish toward some objective and is inspired from nature. It is parallel and random search algorithms.

Unlike lions and monkeys, in animal species such as fish, there is no leader and each member has a self-organizing behaviours. Fish have no knowledge regarding their group and environment and randomly move in the environment via data exchange by their adjacent members. This interaction brings more sophistication to fish swarm optimizaiton.

AFSA has many characteristics, such as fault tolerance and flexibility. It has various applications for problems such as: resource leveling, fuzzy clustering, data mining, spread spectrum code estimation, optimization of the DNA encoding sequence, signal processing, image processing, improving neural networks, job scheduling, and so on. AFSA has many advantages such as high convergence speed and high efficiency, and at the same time it has disadvantages like complexity in time, some imbalance between global and local search.

7.5.2 Artificial Fish Swarm Algorithm

Artificial fish (AF) is a replica of true fish, used to perform analysis and problem explanation Neshat, Sepidnam, Sargolzaei, and Toosi (2012). Fish mostly live in areas where there is availability of ample of food, and they move toward the area with more food by following other fish or by searching individually. The area with a maximum number of fish is generally the most nutritious. Each artificial fish's next behaviour depends upon its current state as well as the local environment's state. An AF influences the environment by its own behaviour as well its companion's behaviour

The AF model contains two parts: variables and functions.

The variables include:

- $X = (x_1, x_2, ..., x_n)$: Current position of each AF.
- *Visual*: Visual distance.
- X_v: State at visual position at some movement.
- *Step*: Step length

- *try_num*: The try number.
- δ: Crowd factor $(0 < \delta < 1)$
- θ: Crowd parameter $\theta \in \{0, 1\}$
- *n*: Number of points (total fish number).

The functions include the behaviours of artificial fish:

- AF_Prey
- AF_Swarm
- AF_Follow
- AF_Move
- AF_Leap
- AF_Evaluate

Figure 7.7 shows the vision of an Artificial Fish.

Behavioural pattern exhibited by fish are:

1. Fish mostly live in an area where ample food is available.
2. The behaviours of fish based on this characteristic are simulated to find the global optimum, the basic idea that works as a background for AFSA.
3. The function Rand () produces a random number between 0 and 1.

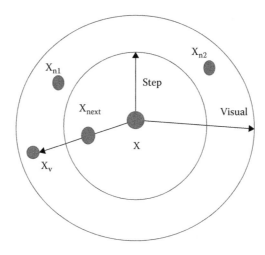

FIGURE 7.7
Visual concept of artificial fish.

4. **AF_Prey**: This is basic biological behaviour of fish with regard to the food. Generally, the fish perceives the concentration of food in water visually to determine where to move, and then chooses the direction.

5. **AF_Swarm:** The fish assemble in groups naturally in the moving process, which is a kind of living habit to guarantee the existence of the colony and avoid dangers.

6. **AF_Follow:** In the moving process of the fish swarm, when a single fish or several ones find food, the neighbourhood partners will move and reach the food quickly.

7. **AF_Move:** Fish swim randomly, and they seek food or companions in larger ranges.

8. **AF_Leap:** Fish stop somewhere; every AF behaviour will gradually be the same and the difference among objective values (food concentration) becomes smaller within some iterations. It might fall into a local extremum and change the parameters randomly to the still states for leaping out of current state.

9. **"Stagnation"** occurs when the objective function value does not change for a certain number of iterations.

```
Start
        for each Artificial Fish AF ∈ [1 ... f]
               initialize Xi
        End
           Blackboard argmin f (Xi)
        for each Artificial Fish AF ∈ [1 ... f]
               Perform Swarm Behaviour on Xi (t) Compute Xi, swarm
               Perform Follow Behaviour on Xi (t) and Compute Xi, follow
        If   f (Xi, swarm) ≥ f (Xi, follow)
               then Xi (t+1) = Xi, follow
        else
               Xi (t+1) = Xi, swarm
        End
   End
```

Pseudo-code for AFSA

7.5.3 Fish Swarm Variants

AFSA is significantly the best swarm intelligence algorithm. However, it has certain disadvantages such as a high time complexity, a lack of balance among global search and local search, and a lack of use of the experience gained from group members for the next moves. Many improvements have been done in this algorithm and many variants have been discovered that contribute to improve the performance of AFSA in some way.

7.5.3.1 Simplified Binary Artificial Fish Swarm Algorithm

A binary version of the artificial fish swarm algorithm is denoted by b-AFSA Azad, Rocha, and Fernandes (2012). In b-AFSA, a decoding algorithm is used to make infeasible solutions feasible.

> **Algorithm Binary Version of AFSA**
> **Begin**
> **Initialize** parameter values
> **Set** k = 1 and randomly initialize xi, i = 1,2,...,I
> decoding is done in order to deal with constraints and evaluate Z.
> Identify xmax and Z(xmax)
> if stop criteria are met then
>   Stop
> **end**
>   **for** each of xi
>   Calculate "visual scope" and "crowding factor"
>   Perform the fish behaviors to create trial point
>   Perform decoding to make the trial point feasible
>   **end**
>   Perform selection to create new current points
>   Evaluate z and identify xmax and z(xmax)
>   **if** k % L = 0
>   Perform leap
>   **end**
>   **Set** k = k+1 && go to **step 4**
> **End**

The b-AFSA algorithm has applications in solving problems such as: the multi-dimensional 0–1 knapsack problem, 0–1 quadratic knapsack problem, facility location problems, etc. Azad, Rocha, and Fernandes (2012), Azad, Rocha, and Fernandes (2014).

7.5.3.2 Fast Artificial Fish Swarm Algorithm

Optimization problems like those addressed by the fish swarm optimizaiton in real life are often highly nonlinear. Global optimization algorithms must be used to in order to obtain optimal solutions. In the AFSA, there are a number of parameters that might have an impact on the final optimization result. Considering the parameter, *Rand ()*, Brownian motion and Lévy flight algorithms (El-Bayoumy et al. 2013) can be used (El-Bayoumy, Rashad, Elsoud, & El-Dosuky, 2014).

The probability density function of the Lévy distribution over the domain $X \ge \mu$ may be defined as:

$$f(x; \mu, c) = \sqrt{\frac{c}{2\Pi}} \frac{e^{-\frac{c}{2(x-\mu)}}}{(x-\mu)^{\frac{s}{2}}} \qquad (7.5.3.2.1)$$

where, μ is location parameter and c is the scale parameter.
The cumulative distribution function is

$$f(x; \mu, c) = ercf\left(\sqrt{\frac{c}{2(x-\mu)}}\right) \qquad (7.5.3.2.2)$$

where, $ercf$ (z) is the complementary error function. The shift parameter μ has the effect of shifting the curve to the right by an amount μ and changing the support to the interval [μ, ∞).

$$f(x; \mu, c)dx = f(y; 0, 1)dy \qquad (7.5.3.2.3)$$

The Lévy distribution has a standard form f $(x; 0, 1)$, which has the following property:

$$\text{where }, y = \frac{x-\mu}{c}$$

Merging the Lévy Flight algorithm for randomization is the major task; a flowchart is shown in Figure 7.8 that is used by the fast artificial fish swarm algorithm El-Bayoumy et al. (2014).

7.5.3.3 Quantum AFSA

AFSA is generally designed to calculate a single optimal solution for a given problem. But in certain practical applications, the global optimal solution as well as some nearby optimal solutions are required. The quantum mechanism is introduced into the AFSA to increase the diversity of species (Zhu, & Jiang, 2010; Zhu, Jiang, & Cheng, 2010). The concepts and principles of quantum computing, such as the quantum bit and quantum gate, are used for modifying the global search ability and the convergence speed of the AFSA, leading to the quantum artificial fish swarm Algorithm (QAFSA) (Zhu, & Jiang, 2010). QAFSA has shown improvement to the various drawbacks of AFSA such as the convergence speed of the AFSA becomes slow at the later stage of the algorithm. The AF may sink into the local optimum and the algorithm cannot get the global optimum for some complicated optimization problems.

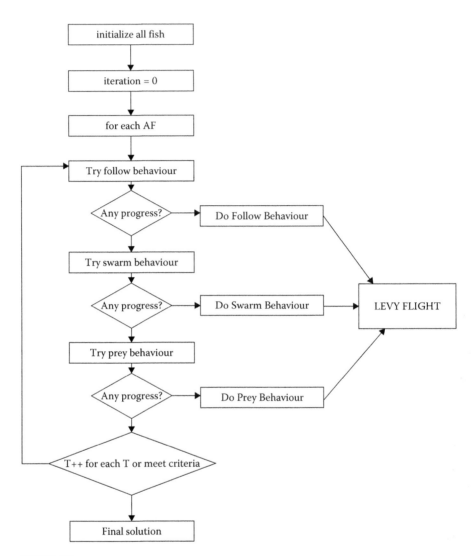

FIGURE 7.8
Fast Artificial Fish Swarm Algorithm.

QAFSA Algorithm

Step 1: Initialize the variables like AF, AFPos, step, visial, try.num, crowd factor δ, mutation probability pα.

Step 2: The solution space transformation for every AF is performed; a fitness value of AF is calculated, and the best AF is included in the bulletin board.

Step 3: AF.Prey, AF.Swarm, AF.Follow, AF.Move are executed by AF and their resultant behaviour is evaluated. Determine the target position and change AF position using Quantum Rotation Gate.

Step 4: Perform mutation operation. Generate Random Number rand between 0 and 1 for every AF, if randl < pα, then execute the mutation operation upon that AF.

Step 5: Perform the Solution Space transformation for each AF, calculate the fitness value of AF, and update the bulletin board.

Step 6: Stop and display the output once the stop criteria are satisfied; otherwise, go to step 3.

7.5.3.4 Chaotic AFSA (CSAFSA)

The AFSA algorithm might get trapped in local optimum in the later period of evolution. Therefore, an artificial fish swarm algorithm based on chaos search (CS) is employed, which is not only able to overcome the disadvantage of easily getting into the local optimum in the later evolution period, but also keeps the rapidity of the previous period. The CS has the advantage of ergodicity, randomness, regularity, and so on, which can ergodically process all states within a certain scope according to their own rules without repetition, thus avoiding local minimum points in the search process, Ma and Wang, (2009). An artificial fish swarm algorithm based on the chaos search (CSAFSA) optimization method makes full use of better global convergence performance, the faster convergence speed of AFSA, and ergodicity of CS. In the search process, AFSA is first used to obtain the neighbourhood of the global optimum solution, and then a secondary search is executed around it.

In CSAFSA (Zhu & Jiang, 2009), in addition to the simple fish behaviours such as AF_Prey, AF_Swarm, AF_Follow, AF_Move, AF_Leap, a bulletin board is maintained. The bulletin board is created to record various states of best AF along with its food concentration. The current state of the AF after each behaviour is compared with the bulletin board and the better state is recorded (Ma & Wang, 2009).

The CS mechanism is brought into the operational flow of AFSA to create CSAFSA. It enhances the global search capabilities, overcoming the local optimum on one hand, and on the other hand, it doesn't reduce the convergence speed and search accuracy. When AF completes a single movement, the global best fish is evaluated and chaos optimization is applied in order to determine the best fish position in a certain radius. If a best is found, the global best fish replaces the solution.

When all of the AF has completed one movement, the global best fish is evaluated, and then the chaos optimization algorithm is used to search around the position of the best fish within a certain radius. If a better one is found, then replace the global best fish with this solution. The CSAFSA algorithm is demonstrated below:

CSAFA Algorithm

Step 1: Initial fish Swarm is randomly generated in the search space.

Step 2: The value in the bulletin board is initialized and the current function value y is calculated for each AF, and the best fish is assigned to the bulletin board.

Step 3: Run a simulation of fish following behaviour and fish swarming behaviour and select the behaviour that gives better function value y; the default behaviour is fish preying.

Step 4: Compare the function value y with the value in the bulletin board, and update the bulletin board with the better value.

Step 5: The chaos search is performed near the current best AF. Replace the global best fish with the better solution found.

Step 6: Judge if preset maximum iteration is achieved or a satisfactory optimum solution is obtained. Otherwise, go to Step 3 or Step 7.

Step 7: Output the optimum solution.

CSAFSA has application in solving NP-hard problems such as Planar Location problem Chen and Tian, (2010).

7.5.3.5 Parallel AFSA

Parallel AFSA can be applied on a graphics processing unit (GPU). The modern GPU has high-end computational power and high memory via parallel, multi-threaded, and multi-core processors (Hu, Yu, Ma, & Chen, 2012; Jiang & Yuan, 2012).

AFSA can be applied on a GPU in parallel, thus giving rise to GPU-based AFSA. This increases the speed of AFSA. The algorithm provides an efficient solution with regard to handling issues of a big swarm population with a large dimension.

GPU-AFSA Algorithm
Initialize the states of AFs randomly
Initialize the Random Number Array
Initialize AFSA parameters
Transfer CPU data to GPU
$Y_{min} \leftarrow +\infty$
for iter = iter to iterNum
    Calculate fitness value of all AFs: TESTF
    Update the optimization result: Y_{min}
    Calculate distance between AFs: DIST
    Execute fish preying process
    Execute fish swarming process
    Execute fish following process
    Update the state of each AF: X
Transfer Y_{min} back to CPU and output

7.6 Conclusion

To tackle complex real world problems, researchers have been looking into natural processes and creatures – both as model and metaphor – for years. This chapter gives a comprehensive study on four population based stochastic search algorithms, namely, the Cuckoo Search, Glowworm, WASP and Fish Swarm optimization algorithms. These algorithms are designed to essentially solve the constraint handling, "stuck in local optima" problems in the real-value domain. All these algorithms have demonstrated their potential to solve many optimization problems as they have been applied to a variety of problems, ranging from scientific research to industry and commerce.

References

AwadEl-bayoumy, M. A., Rashad, M.Z., Elsoud, M. A., & El-Dosuky, M. A. (2013). *FAFSA: fast Artificial Fish Swarm Algorithm, International Journal of Information Science and Intelligent System* 2(4) 60–70.

Azad, M. A. K., Rocha, A. M. A., & Fernandes, E. M. (2012 June). Solving multi-dimensional 0–1 knapsack problem with an artificial fish swarm algorithm. In *International Conference on Computational Science and Its Applications* (pp. 72–86). Berlin: Springer.

Azad, M. A. K., Rocha, A.M.A., & Fernandes, E.M. (2014). A simplified binary artificial fish swarm algorithm for 0–1 quadratic knapsack problems, *Journal of Computational and Applied Mathematics* 259 897–904.

Bao, L. & Zeng, J.-C. (2009). Comparison and analysis of the selection mechanism in the artificial bee colony algorithm. In *Ninth International Conference on Hybrid Intelligent Systems 2009* (pp. 411–416).

Beckers, R., Deneubourg, J. L., Goss, S., & Pasteels, J. M. (1990). Collective decision making through food recruitment. *Insectes Sociaux, Paris* 37(3): 258–267. Springer.

Bhandari, A. K., Singh, V. K., Kumar, A., & Singh, G. K. (2014a). Cuckoo search algorithm and wind driven optimization-based study of satellite image segmentation for multilevel thresholding using kapur's entropy, *Expert Systems with Applications* 41(7), 3538–3560.

Bhandari, A., Soni, V., Kumar, A., & Singh, G. (2014b). Cuckoo search algorithm-based satellite image contrast and brightness enhancement using dwt-svd, *ISA Transactions* 53(4), 1286–1296.

Bonabeau, E., Dorigo, E., & Theraulaz, G. (1999). Swarm Intelligence: From Natural to Artificial Systems. Oxford, UK: Oxford University Press, pp. 1–278.

Cao, Y., Yang, Y., & Wang, H. (2008). Integrated routing Wasp algorithm and scheduling wasp algorithm for job shop dynamic scheduling. In *International Symposium on Electronic Commerce and Security 2008* (pp. 674–678).

Chen, Z. & Tian, X. (2010, March). Notice of retraction artificial fish-swarm algorithm with chaos and its application. In *2010 Second International Workshop on Education Technology and Computer Science (ETCS)*. IEEE (Vol. 1, pp.226–229).

Cicirello, V. A. & Smith, S. F. (2004). Wasp-like agents for distributed factory coordination, *utonomous Agents and Multi-Agent Systems 8*(3) 237–267.

Cicirello, V. A., & Smith, S. F. (2002, January). Distributed coordination of resources via wasp-like agents. In *First NASA GSFC/JPL Workshop on Radical Agent Concepts (WRAC).*

Cicirello, V. A., & Smith, S. F. (2001). Wasp nests for self-configurable factories Agents. In *Proceedings of the Fifth International Conference on Autonomous Agents May–June 2001,* ACM Press.

Cobos, C., Muñoz-Collazos, H., Urbano-Munoz, R., Mendoza, M., León, E., & Herrera-Viedma, E. (2014). Clustering of web search results based on the cuckoo search algorithm and balanced Bayesian information criterion *Information Sciences 281* 248–264.

Dehuri, S., Sung-Bae, Cho, & Ashish, Ghosh. (2008). WASP: A multi-agent system for multiple recommendations problem. *4th International Conference on Next Generation Web Services Practices, 2008. NWESP'08.* IEEE

El-Bayoumy, M. A., Rashad, M. Z., Elsoud, M. A., & El-Dosuky, M. A. (2014). Job scheduling in grid computing with fast artificial fish swarm algorithm. *International Journal of Computer Applications 96*(14), 1–5

Feng, D., Ruan, Q, & Limin, D. (2013). Binary cuckoo search algorithm. *Jisuanji Yingyong/Journal of Computer Applications 33*(6), 1566–1570.

Gherboudj, A., Abdesslem Layeb, & Salim, Chikhi. (2012). Solving 0-1 knapsack problems by a discrete binary version of cuckoo search algorithm. *International Journal of Bio-Inspired Computation 4*(4), 229–236.

Giveki, D., Salimi, H., Bahmanyar, G., & Khademian, Y. (2012). Automatic detection of diabetes diagnosis using feature weighted support vector machines based on mutual information and modified cuckoo search. *The Computing Research Repository (CoRR), ACM.*

Goel, S., Sharma, A., & Bedi, P. (2011). Cuckoo search clustering algorithm: A novel strategy of biomimicry. In *2011 World Congress on Information and Communication Technologies (WICT)* (pp. 916–921). IEEE.

Hu, Y., Yu B., Ma, J., & Chen, T. (2012, May). Parallel fish swarm algorithm based on GPU-acceleration. In *2011 3rd International Workshop on Intelligent Systems and Applications (ISA)* (pp. 1–4). IEEE.

Iourinski, D., Starks, S. A., Kreinovich, V., & Smith, S. F. (2002). Swarm intelligence: theoretical proof that empirical techniques are optimal. In *Automation Congress, 2002 Proceedings of the 5th Biannual World* (Vol. 13, pp. 107–112). IEEE.

Jiang, M., & Yuan D. (2012). Parallel artificial fish swarm algorithm. In *Advances in Control and Communication.* Berlin: Springer, pp. 581–589.

Kaipa, K. N., & Ghose, D. (2017). *Glowworm Swarm Optimization: Theory, Algorithm and Applications,* Berlin: Springer International Publishing AG.

Karaboga, D., & Akay, B. (2009). A comparative study of artificial bee colony algorithm, *Applied Mathematics and Computation 214,* 108–132.

Krishnanand, K. N., & Ghose, D. (2005). "Detection of multiple source locations using a glowworm metaphor with applications to collective robotics." *Swarm Intelligence Symposium.* SIS 2005 Proceedings, IEEE.

Litte, M. (1977). Behavioral ecology of the social wasp, *Mischocyttarus Mexicanus. Behavioral Ecology and Sociobiology 2* 229–246.

Liu, X., & Fu, H. (2014). PSO-based support vector machine with cuckoo search technique for clinical disease diagnoses. *The Scientific World Journal 2014*, 548483.

Ma, H., & Wang, Y. (2009, August). An artificial fish swarm algorithm based on chaos search. In *ICNC'09. Fifth International Conference on Natural Computation*, IEEE (Vol. 4, pp. 118–121).

Meziane, R., Boufala, S., Amar, H., & Amara M. (2015). Wind farm reliability optimization using cuckoo search algorithm. *Recent Researches in Electrical Engineering*, pp 295–303.

Neshat, M., Sepidnam, G., Sargolzaei, M., & Toosi A. N. (2012). Artificial Fish Swarm Algorithm: A Survey of the State-of-the-art, Hybridization, Combinatorial and Indicative Applications. *Artificial Intelligence Review*, 1–33. DOI 10.1007/s10462-012-9342-2

Ouaarab, A., Ahiod, B., & Yang, X.-S. (2014). Discrete cuckoo search algorithm for the travelling salesman problem. *Neural Computing and Applications 24*(7–8), 1659–1669.

Ouyang, X., et al. (2013). A novel discrete cuckoo search algorithm for spherical travelling salesman problem. *Applied Mathematics & Information Sciences 7*(2), 777.

Pereira, L. A. M. et al. (2014). *Cuckoo Search and Firefly Algorithm. In A Binary Cuckoo Search and its Application for Feature Selection*. Cham: Springer, pp. 141–154.

Pinto, P., Runkler, T. A., and Sousa, J. M. (2005). Wasp swarm optimization of logistic systems. In Ribeiro, B., Albrecht, R. F., Dobnikar, A., Pearson, D. W., and Steele, N. C. (eds) *Adaptive and Natural Computing Algorithms*. Vienna: Springer.

Raja, N. S. M., & Vishnupriya, R. (2016). Kapur's entropy and cuckoo search algorithm assisted segmentation and analysis of RGB images. *Indian Journal of Science and Technology 9*(17).

Runkler, T. A. (2008). Wasp wwarm optimization of the c-means clustering model. *International Journal of Intelligent Systems 23*, 269–285.

Soto R. et al. (2015). A binary cuckoo search algorithm for solving the set covering problem. In *International Work-Conference on the Interplay between Natural and Artificial Computation*. Cham: Springer.

Simian, D., Simian, C., Moisil, I., & Pah, I. (2007). Computer mediated communication and collaboration in a virtual learning environment based on a multi-agent system with wasp-like behavior. In *Lecture notes in Computer Science* (LSCC 2007). Berlin: Springer, pp. 606–614.

Simian, D., Stoica, F., & Curtean-Baduc, A. (2008). Multi-Agent System models for monitoring optimization within a site, Natura 2000, Mathematics and Computers in Biology and Chemistry. In *Proceedings of 9th WSEAS International Conference on Mathematics and Computers in Biology and Chemistry MCBC'08*, Bucharest, Romania: WSEAS Press, pp. 212–217.

Simian, D., Stoica, F., & Simian, C. (2010). Optimization of complex SVM kernels using a hybrid algorithm based on wasp behavior. *Lecture notes in computer science*, LNCS 5910, pp. 361–368. 10.1007/978-3-642-12535-5_42.

Srivastava, P. R., Chis, M., Deb, S., & Yang, X. S. (2012). An efficient optimization algorithm for structural software testing. *Int J. Artificial Intelligence 9*(S12), 68–77.

Theraulaz, G., Bonabeau, E., & Deneubourg, J. L. (1995). Self-organization of hierarchies in animal societies: the case of the primitively eusocial wasp *polistesdominuluschrist. Journal of Theoretical Biology 174*, 313–323.

Theraulaz, G., Bonabeau, E., & Deneubourg, J. L. (1998). Response threshold reinforcement and division of labour in insect societies. *Proceedings of the Royal Society of London* B 265(1393), 327–335.

Theraulaz, G., Goss, S., Gervet, J., & Deneubourg, J. L. (1991). Task differentiation in polistes wasp colonies: A model for self-organizing groups of robots. In *From Animals to Animats. In Proceedings of the First International Conference on Simulation of Adaptive Behavior*. Cambridge, MA: MIT Press, pp. 346–355.

Wang, G.-G. et al. (2016). Chaotic cuckoo search. *Soft Computing 20*(9), 3349–3362.

Tzy-Luen, N., Keat, Y. T., & Abdullah, R. (2016). Parallel Cuckoo Search algorithm on OpenMP for travelling salesman problem. In *3rd International Conference on IEEE, Computer and Information Sciences (ICCOINS)* (pp. 380–385). IEEE.

Yang, X. S., ed. *Cuckoo Search and Firefly Algorithm: Theory and Applications* (Vol. 516). Springer, 2013.

Yang, X. S., & Deb, S. (2009). Cuckoo search via Levy Flights. In *Proceedings of World Congress on Nature & Biologically Inspired Computing (NaBIC 2009)*. USA: IEEE Publications, pp. 210–214.

Yang, X.-S., & Deb, S. (2013). Multi-objective cuckoo search for design optimization. *Computers & Operations Research 40*(6), 1616–1624.

Yazdani, D., Toosi, A. N., & Meybodi, M. R. (2010, December). Fuzzy adaptive artificial fish swarm algorithm. In *Australasian Joint Conference on Artificial Intelligence*. Berlin: Springer, pp. 334–343.

Yildiz, A. R. (2013). Cuckoo search algorithm for the selection of optimal machining parameters in milling operations, *The International Journal of Advanced Manufacturing Technology 64*(1-4), 55–61.

Zheng, H. & Zhou, Y. (2012). A novel cuckoo search optimization algorithm based on guass distribution. *Journal of Computational Information Systems 8*(10), 4193–4200.

Zhu, K., & Jiang, M. (2009, December). An improved artificial fish swarm algorithm based on chaotic search and feedback strategy. In *CiSE 2009. International Conference on Computational Intelligence and Software Engineering*. IEEE, pp. 1–4.

Zhu, K., Jiang, M., & Cheng Y. (2010, October). Niche artificial fish swarm algorithm based on quantum theory. In *IEEE 10th International Conference on Signal Processing (ICSP)* (pp. 1425–1428). IEEE.

Zhu, K., & Jiang, M. (2010, July). Quantum artificial fish swarm algorithm. In *2010 8th World Congress on Intelligent Control and Automation (WCICA)* (pp.1–5). IEEE.

Zaw, M. M. and Mon, E. E. (2013). Web document clustering using cuckoo search clustering algorithm based on levy flight, *International Journal of Innovation and Applied Studies 4*(1): 182–188.

8

Misc. Swarm Intelligence Techniques

M. Balamurugan, S. Narendiran, and Sarat Kumar Sahoo

CONTENTS

8.1 Introduction

Optimization is an effective tool to determine the best input while obtaining the output with minimal cost. The optimization problem process is started by selecting the appropriate variable and by framing the limits and fitness functions. The key aim of the fitness function is to find the optimum response by selecting the appropriate variables with respect to the provided constraints. There are different types of optimization methods to resolve the particular variable or multi-variable functions with or without limits. The optimization methods are broadly classified as traditional method and non-traditional methods or population-based methods (Bonabeau, Dorigo, & Theraulaz, 1999).

The traditional methods are more appropriate for the problems that have a single objective function, but they are not effective for multi-objective and complex optimization problems. The population-based techniques are more applicable to difficult problems. The non-traditional methods are further classified into evolutionary techniques, nature-based techniques, and swarm intelligence methods and logical techniques.

Evolutionary computation primarily depends on a calculation system with the concept of computer-generated evolution. Evolutionary computation contains of a constant or adjustable-size population, an objective function, and a genetically inspired operator to progress through the subsequent generations of the current generation. Samples of evolutionary systems are differential evolution and genetic algorithms.

Logic-based methods are developed based on the combination of mathematical and logical programs. The tabu search optimization method is the example of a logic-based method (Hooker, 1994).

Swarm intelligence is based on the concept of artificial intelligence as the combined behaviour of systems in natural or simulated environment. Gerardo Beni and Jing Wang introduced the concept of swarm intelligence in 1989, based on cellular robotic systems. Generally, a swarm intelligence system consists of a group of simple representatives that act together in a neighbourhood with one another and with their location. The swarm intelligence algorithms are framed based on the inspiration from systems, specifically biological systems. Usually, the individual agents interact with the other agents locally to find the global behaviour of the agents. The examples of such naturally inspired algorithms are bird flocking, bacterial growth, and ant colonies, which are mostly referred to as the general set of algorithms (Parpinelli & Lopes, 2011).

The properties of typical swarm intelligence systems are:

- Generally, they consist of many individuals.
- The individuals are identical or belong to a small number of topologies.
- The relations between the entities are established by using certain rules, which exploit the local information that can be exchanged between the individuals directly or through the location.
- The self-organized behaviour of the scheme is characterized as the overall behaviour of the scheme as a consequence of the collaboration between individuals.

Miscellaneous swarm intelligence techniques report the study of the combined performance of the systems provided by several mechanisms that have been organized by using distributed controls, and the individual system mechanism has been demonstrated in the functioning principles of communal insects and animals such as ants, fish, etc. The analytical-based method is developed by linking the logical and mathematical programming. To determine the optimal solutions for a given problem, the structure has been extracted from the mathematical programs, and the information has been extracted from the logical programs.

The chapter will discuss various SI techniques such as the Termite Hill Algorithm, Cockroach Swarm Optimization, the Bumblebee Algorithm, the Monkey Search Algorithm, the Social Spider Optimization Algorithm, Cat Swarm Optimization, Intelligent Water Drop, Dolphin Echolocation, biogeography-based optimization, the Paddy Field Algorithm, the Weightless Swarm Algorithm and Eagle Strategy along with their working methodologies and mathematical proof.

8.2 Termite Hill Algorithm

Termite Hill algorithm is the probabilistic and non-demand routing algorithm especially used in the area of hills to discover the most appropriate path by utilizing the behaviour of real termites. The Termite Hill algorithm is applicable for wireless sensor networks because it has been motivated by termites. To follow emergent routing behaviour, a swarm intelligence principle has been used for each packet to define new rules. This algorithm has shown better performance because of quick route discovery; for route selection criteria, existing energy has been utilized, control traffic has been reduced, and memory usage has been reduced along with additional benefits. Each node of this algorithm works as source and router and the hill is the specific node named as sink, and the number of sinks has been decided according to the network size, which is similar to the termite ad hoc network (Zungeru, Ang, & Seng, 2012).

Pheromone table and the route selection are the two processes used to design the Termite Hill algorithm.

8.2.1 Pheromone Table

The data collected by the forward soldier has been placed in the pheromone table. The volume of pheromone on every local path has been maintained in the table by each node. $T_{n,d}$ is represented as the record in the pheromone table, where m is the neighbour index and c is the dimension. When the pheromone table is incremented by the parameter γ when a source s is transported to the previous destination hop r, the increment is updated by Equations (8.1) and (8.2).

$$T'_{r,s} = T_{r,s} + \gamma \tag{8.1}$$

$$\gamma = \frac{M}{V - \left(\dfrac{V_{min} - M_j}{V_{av} - M_j}\right)} \tag{8.2}$$

where γ is denoted as reward, V is the initial energy, V_{av} & V_{min} are the average and minimum energy, N_j is the number of active nodes that have been visited by the forward soldier, and M is total number of nodes. Evaporation factor e^{-v} has been multiplied with every record in the pheromone table. The pheromone decay has been calculated by Equation (8.3).

$$T'_{m,c} = T'_{m,c} * e^{-v} \tag{8.3}$$

Equation (8.3) with a slow decay rate can be applied for some applications with flexibility and robustness. To calculate better pheromone decay, a percentage of its own value has been subtracted by using Equation (8.4).

$$T'_{m,c} = (1 - x)T'_{m,c} \tag{8.4}$$

8.2.2 Route Selection

The distribution of uniform probability for each routing node has been calculated by using Equation (8.5).

$$P_{r,d} = \frac{1}{N} \tag{8.5}$$

where $P_{r,d}$ is the probability of node from source to endpoint. If the packet has to be forwarded to the nearest neighbour then the probability distribution parameter has been adjusted by using the Equation (8.6).

$$P_{r,d} = \frac{(T_{s,d} + \alpha)^{\beta}}{\sum_{i=1}^{N}(T_{i,d} + \alpha)^{\beta}} \tag{8.6}$$

The constraints α and β are used to adjust the algorithm.

8.2.3 Analogy of Termites

The termite uses the following process to search the food by laying chemical trails along with abdominal sterna gland secretions. While searching for food, the termites walk slowly, laying the trail, and once food is found, they come back at rapid speed to the nest, recruiting the nestmates to hunt for the food by secreting chemicals from the abdominal gland. If there are no obstacles, the termites can rapidly find the food source and bring back the food efficiently. If the termite faces obstacles, it can readily choose the shortest path to find the food.

Termites can search for their food by discovering routes, if they remain vital. When a node takes certain records, these remain communicated to sink node; by generating the information from the forward soldier, it can display this to its neighbours. When the middle node is received, then it searches for a legal way to find its endpoint, and if the search is positive, then the rear soldier has been generated, and it again sends the message to the origin point. If it finds a valid route, then the termite can reach its origin point; otherwise, it can search for next entry and the process is repeated. The flowchart of the Termite Hill Algorithm is presented in Figure 8.1.

In the development of the algorithm strategy, the subsequent assumptions were also prepared:

1. In the network every node has been linked to one or more neighbours.
2. A node can act as a router for communication between one node and other nodes and it may also act as a source as well as destination.
3. Before making a decision it knows the information about the neighbors or network configuration.
4. The same amount of energy is required to send the message to any node throughout the network.

The steps of the Termite Hill algorithm are described below.
Step 1: The termites on the node are initialized.
Step 2: The pheromone is updated by using Equations (8.1) and (8.2).
Step 3: Route selection is performed by using Equation (8.5).
Step 4: Redo the steps 2–3 until the end criteria are met.

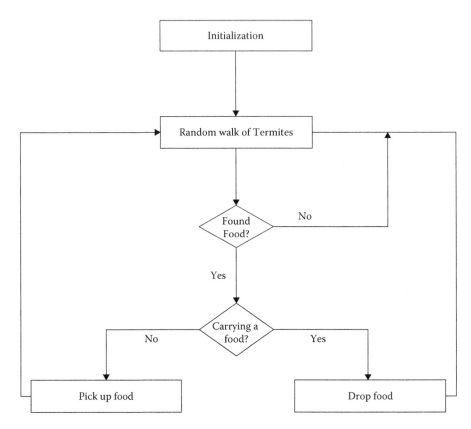

FIGURE 8.1
Flowchart of Termite Hill Algorithm.

8.3 Cockroach Swarm Optimization

Cockroach Swarm Optimization (CSO) has been formulated as a result of the inspiration of cockroach swarm hunting behaviour, and this algorithm has been built by imitation of the chase-swarming behaviour of individual cockroaches seeking to find the overall best value. Cockroaches are insects and are mostly seen in the places that provide dark, worm, moist shelter. They exhibit food hunting behaviour like chasing, swarming, and dispersing (Kwiecien & Pasieka, 2017).

The CSO algorithm has been implemented by mimicking the biological behaviours of cockroaches: chase-swarming, dispersing, and ruthless behavior. The modeling equation of each process is mentioned below.

8.3.1 Chase-Swarming Behaviour

$$y_r = \begin{cases} y_r + a * rand * (\rho_r - y_r), \ y_r \neq \rho_r \\ y_r + a * rand * (\rho_g - y_r), \ y_r \neq \rho_r \end{cases} \tag{8.7}$$

where y_r is the position of cockroach, a is the fixed value, *rand* is the arbitrary value between (0, 1), ρ_r is the individual best, and ρ_g is the global best cockroach point, and the personal best can be calculated by using Equations (8.7) and (8.8).

$$\rho_r = opt_s\{y_s, |y_r - y_s| \leq visual\} \tag{8.8}$$

where visual is constant, $r = 1,2,3,...N$, $s = 1,2,3,...N$, and the global best position has been determined by using Equation (8.9).

$$\rho_z = opt_r\{y_r\} \tag{8.9}$$

8.3.2 Dispersing Behaviour

$$y_r = y_r + rand(1, E), r = 1, 2, ..., N \tag{8.10}$$

where E is a dimensional arbitrary position that can be fixed with certain range.

8.3.3 Ruthless Behaviour

$$y_l = \rho_z \tag{8.11}$$

where l is an arbitrary integer between [1, N].

The basic steps of the CSO algorithm are described below.

Step 1: Initialize the algorithm constraints by means of dimension, step, rand, and visual.
Step 2: Search the local and global optimum position ρ_i and ρ_z.
Step 3: Implement the chase-swarming behaviour, and the global optimum is updated.
Step 4: Implement the dispersing behaviour, and the global optimum is updated.
Step 5: Implement ruthless behaviour.
Step 6: Redo the steps 2–5 until the end criteria are met.

8.3.4 Cockroach Swarm Optimization Procedure for the Classic Travelling Salesman Problem

Step 1: Initialize the parameters and swarm size of the algorithm; a represents population size, and b represents food size. Optimal food is chosen as the local optima food, and it is represented by z.

Step 2: All the cockroaches in the search space crawl to food
        For (int k=1; k<=b; k++)
        For (int k=1; k<=a; k++)
        {

                {Compute the solution when cockroach crawl to food
                If (solution is better than z)
                        {The solution has been updated through z}

                }
        Then new solution has been set to cockroach

        }

Step 3: All cockroaches crawl to z
        For (int k=1; i<=b; k++)
        {

                {Compute the solution when cockroaches crawl to z
                If (solution is better than z)
                        {The solution has been updated through z}

                }
        Then new solution has been set to cockroach

        }

Step 4: The iteration is stopped when termination criteria are met and the output is displayed; otherwise, go to step 2.

8.4 Bumblebee Algorithm

The Bumblebee algorithm is formulated based on inspiration from the breeding performance of bumblebees. To estimate the value of the solution, a fitness function has been framed in most of the optimization algorithms. But in the Bumblebee algorithm, the fitness function is used to determine the significance of the solution along with the lifespan of each individual in a group in the bumblebee colony (Marinakis, Marinaki, & Matsatsinis, 2010).

Let's discuss the implementation of bumblebee's algorithm for the classical k-colouring of a graph problem in graph theory. The pseudo-code for the bumblebee's algorithm has been mentioned below.

8.4.1 Pseudo-Code of Bumblebee Algorithm

Start
Set arbitrary solutions;
      Make S arbitrary solutions of the problem;
Initialize Maximum Generations,
Assign a x b cells world;
      Create the colony nest with S bumblebees;
      Put E food cells at random;
      Assign a random solution to each bumblebee;
      Assign the lifespan of each bumblebee according to its solution
         fitness;
Redo Until (Current generation > Maximum generation) Do
      Look for the most suitable bumblebee;
      If (its solution is the global optimum) Then
           Inform the solution and end the algorithm;
      End if
      Reduce life counters;
      Gain bumblebees;
      Generate a fresh bumblebee in the nest
         Each H generation;
      Move arbitrarily each single bumblebee to a neighbour cell;
      If (a bumblebee discovers food) Then
         Reduce food counter;
         Rise lifetime of the bumblebee;
         Move bumblebee rear to the nest;
      End if
      Mutate bumblebee solutions;
      Again, calculate fitness then allocate new lifetimes;
      Increase Current generation;
End Do
End.

8.4.2 Working Process of Bumblebee Algorithm

The colony of bumblebees is arranged in the form of square with $n \times m$ cells, where each cell must have any one of the following contents, which are a bumblebee or food or empty, or it has nest and this arrangement is called artificial work, which is an essential part of determining the lifespan of a bee.

The bumblebees are present in the nest during the initial stage. When the bees started to come out from the nest, the position of each bee is chosen randomly within the search space of $n \times m$ cells. When a bumblebee finds food, it immediately comes back to its nest, increases its lifespan by two years, and saves the position of food for the next bumblebee. The bumblebee comes out from the nest only when another bumblebee super-sedes it. The counter, which contains food, has the counter value of 20 during the start of the phase, and it is reduced by one unit during each iteration. When the counter value touches zero, a fresh food counter is produced in an arbitrary position.

The bumblebee colony consists of a queen, female workers, and male workers, which are also called drones. The queen is the only member that contains the value of best positions and passes this information to the new born bees every few generations. Local minima are avoided because the process always starts with the best positions.

As has been discussed earlier, the life of the bumblebee is reduced by one unit after each generation. When it reaches zero, it is removed from the world, and it has been known that the convergence size is determined by the lifespan of the bumblebee, so it is very hard to catch the right convergence conditions. In order to resolve this problem, the reaper mechanism is used to select the lifespan of the bumblebee, which improves the algorithm's performance.

The steps of the bumblebee algorithm are defined below.

Step 1: Initialize a square grid in the form of toroid with a x b cells with four probable states allocated to every cell, which are categorized as containing a nest or with bumblebee, with food, or empty.

Step 2: At the start, all the bumblebees are placed in the nest. At each generation, every individual bumblebee will fly out of the nest one by one arbitrarily to a farther position in the nest around its current position.

Step 3: Every bumblebee is allocated an arbitrarily produced solution at the initial stage. Some of the best solutions are recorded by the queen, who sends the information to the newly generated bumblebee.

Step 4: Similar to genetic algorithm, mutation is presented in this algo-rithm with a slight adjustment while applying the mutation to a bumblebee.

Step 5: A reaper mechanism is employed to avoid over-population in this algorithm.

Therefore, when the age of the bumblebee arrives at zero, the individual bumblebee is removed from the nest.

8.5 Social Spider Optimization Algorithm

The Social Spider Optimization algorithm has been formulated based on the hunting approach of the social spider. The hunt space of the spider has been expressed by means of a hyper-dimensional web. Every point on the spider web denotes the result toward this optimization problem, and it is used as the transmission media for the spider to search for the location of prey by generating vibrations. The spider cannot move out of the web because it contains the solutions that are feasible. Each vibration has certain information about one spider, and it transmits the information to the other spiders on the web (Yu & Li, 2015).

The steps of the social spider algorithm are defined below.

Step 1: To perform optimization, social spider algorithm spiders are measured as mediators, and each has predefined memory to store subsequent material. The fitness function measures the present location of the spider and the target vibration of spider in the previous iteration. Further, it calculates the number of iterations since spider has last changed its target vibration along with the movement that spider performed in the previous iteration. Then, the dimension mask has been employed to guide movement for the spider in the previous iteration.

Step 2: The search space of the spider is defined in the range $[0, +\infty]$. When the spider moves into the different position, this is denoted as $P_a(t)$ with respect to time t. As the spider moves into the new position, this is represented by $I(P_m, P_n, t)$, where the strength of the spider P_n is sensed by the source spider P_m. Therefore, the intensity of the spider can be defined using Equation (8.12).

$$I(P_r, P_r, t) = log\left(\frac{1}{f(P_r) - B} + 1 \right) \tag{8.12}$$

where B is the confidentiality small constant.

The distance between spider m and spider n can be computed with 1-norm Manhattan distance by using the Equation (8.13).

$$D(P_m, P_n) = \|P_m - P_n\|_1 \tag{(8.13)}$$

Σ is denoted as the standard deviation to calculate the attenuation vibration by using the Equation (8.14).

$$I(P_m, P_n, t) = I(P_m, P_m, t) \times exp\left(-\frac{D(P_m, P_n)}{\sigma \times s_a} \right) \tag{8.14}$$

The parameter s_a controls the attenuation level of vibration strength over distance. When s_a is smaller, stronger attenuation is executed on the vibration.

Step 3: When the mask dimensions have been determined, the new position $P_{r,i}^{fo}$ of the spider is calculated by using Equation (8.15).

$$P_{r,i}^{fo} = \begin{cases} P_{r,i}^{tar} \, a_{r,i} = 0 \\ P_{r,i}^{z} \, a_{r,i} = 1 \end{cases} \qquad (8.15)$$

where z is a arbitrary integer between $[1, |pop|]$, and $a_{r,i}$ stand for the ith dimension of mask m. Then the random walk of the spider can be calculated by using the Equation (8.16).

$$P_r(u+1) = P_r + (P_r - P_r(u-1)) \times z + (P_r^{fo} - P_r) \odot T \qquad (8.16)$$

where $\odot$ represents the component-based multiplication, and T is the vector of float point number, which has been generated randomly between $[0,1]$.

Step 4: The last step of the iteration is limitation control. During the movement of a spider from one position to another position, sometimes it may move out from the web, and to address this violation constraint a border-free restriction position $P_r(u+1)$ is calculated by using the Equation (8.17).

$$P_{s,i}(t+1) = \begin{cases} (\bar{x}_i - P_{s,i}) \times r, ifP_{s,i}(t+1) > \bar{x}_i \\ (P_{s,i} - \underline{x}_i) \times r, ifP_{s,i}(t+1) < \underline{x}_i \end{cases} \qquad (8.17)$$

where $\underline{x}_i$ and $\bar{x}_i$ are the lower and higher bounds of the ith dimension in the search area, and r is an integer randomly generated in the range $[0,1]$.

8.5.1 Pseudo-Code for Social Spider Optimization Algorithm

Set values to the variables of Social Spider Algorithm
Generate the population of spider's pop and set remembrance for them
Set u_s^{tar} for every spider
while end condition not met do
      for every spider r in pop do
           compute the fitness assessment of r
           Make a vibration at the point of r
      end for
      for every spider r in pop do
           Compute the strength of the vibrations U produced by each
           spider
           Pick the vibration intensity u_s^{best} from U
           if the strength of u_s^{best} is greater than u_s^{tar} then
               Store u_s^{best} as u_s^{tar}

end if
Update d_s
Produce an arbitrary number s from [0, 1]
If $s > p_b^{b_s}$ then
        dimension mask n has been updated$_s$
end if
Produce P_r^{fo}
Do a random walk
Address some violated conditions
    end for
end while
Display optimum solution.

8.6 Cat Swarm Optimization

Cat Swarm Optimization has been formulated based on the hunting strategy of cat. Generally, cats occupy most of their time in rest and spend little time for hunting and catching prey. Even at rest cats have high alertness of their surroundings and the objects moving in the environment (Bahrami, Bozorg-Haddad, & Chu, 2018).

Based on the hunting strategy of cats, Tsai and Chu developed the cat swarm optimization with two approaches: tracing mode and seeking mode. Firstly, the cats are arbitrarily dispersed all over the search space, then the cats are subdivided into two groups. The first group is called the seeking group, where the cats are in rest and keeping an eye on the environment, and the second group is called the tracing mode, where the cats are chasing prey. The combination of these group supports finding the global optimization value. Usually more cats are placed in the first group because cats spend little time searching for their food. The ratio that is used to divide the cats among the groups is called the mixture ratio.

The computational method of cat swarm optimization is defined below.

Step 1: The preliminary population of cats is created, and they are separated into the dimensional solution area $(X_{i,d})$, then arbitrarily set the velocity for every cat with maximum velocity $(v_{i,d})$.

Step 2: As per the value of mixture ratio, set a flag to every single cat to arrange the cat systematically in tracing mode or seeking mode.

Step 3: Compute the appropriate solution for every cat and store the value of the cat that has the optimum fitness function. X_{best} represents the position of best cat so far.

Step 4: If the termination criteria have been reached, then the process is stopped; otherwise, repeat the steps 2, 3, and 4.

8.6.1 Seeking Mode

The cat swarm optimization algorithm in seeking mode is operated by the aspects counts of dimension to change (CDC), seeking range of selected dimension (SRD), seeking memory pool (SMP), and self-position consideration (SPC). The maximum change between the old and new positions is defined by SRD. The quantity of prints generated for every single cat in seeking mode is defined by SMP. CDC defines the dimensions used for mutation, and the present location of the cat is termed SMP.

The procedure for seeking mode is defined below.

1. Every print of cat is denoted by SMP, and if the SPC is true, then the current position is copied by SMP
2. Calculate the CDC for each copy by using Equation (8.18)

$$X_{cn} = (1 \pm SRD \times R) \times X_c \tag{8.18}$$

where X_c is the current position, X_{cn} is the new position, and R is an arbitrary value in the range [0,1]

3. Calculate the appropriate assessment for the new position. If the fitness values of all cats are equivalent, then assign probability value of 1 for all the positions; otherwise, calculate the probability by using Equation (8.19)

$$P_i = \frac{|FS_i - FS_b|}{|FS_{max} - FS_{min}|}, where\ 0 < i < j \tag{8.19}$$

where P_i is the probability of the current cat, FS_i is the fitness value of cat, FS_{max} and FS_{min} are maximum and minimum value of fitness function, $FS_b = FS_{min}$ for maximization problems and $FS_b = FS_{max}$ for minimization problems.

4. Moving the cat from one position to another position or replace the position of the cat can be done randomly by using a roulette wheel.

8.6.2 Tracing Mode

Tracing mode has been used to help a cat chase prey, and the velocity of the cat is determined by Equation (8.20).

$$u_{j,c} = u_{j,c} + s_1 \times d_1 \left(Y_{best,c} - Y_{j,c} \right) \tag{8.20}$$

where $u_{j,c}$ is the cat's velocity in j dimension, $Y_{best,c}$ is the optimum position of the cat, $Y_{j,c}$ is the present location of the cat, d_1 is constant, s_1 is an arbitrary integer between [0,1]. With the help of velocity, the cat moves its position in the dimensional search space, then saves the new location. If the cat's velocity is greater than maximum velocity, then the value is assigned as maximum velocity. Then the new location of the cat is defined using Equation (8.21).

$$Y_{j,c,new} = Y_{j,c,old} + u_{j,c} \qquad (8.21)$$

Where, $Y_{j,c,new}$ & $Y_{j,c,old}$ are the new and current location of cat dimension d.

8.6.3 Pseudo-Code for Cat Swarm Optimization Algorithm

Start
Input factors of the algorithm and the preliminary data
Set the cat population Xi (i = 1, 2, . . ., n), v and self-position consideration
While (the end condition is not met or I < I_{max})
      Compute the fitness function solution for all cats and arrange them
      X_g = cat with the best solution
            For i = 1: J
                  If self-position consideration = 1
                        Begin seeking mode
                        Else
                        Begin tracing mode
                End if
            End for i
      End while
      Post processing the parameters then display the results
End

8.7 Monkey Search Algorithm

Monkey Search algorithm is a population-based swarm intelligence algorithm that has been developed in recent times based on the monkey climbing process in the mountains. The major steps followed in this algorithm are climb process, watch-jump process, and somersault process (Marichelvam, Tosun, & Geetha, 2017).

8.7.1 Computational Procedure of Monkey Search Algorithm

1. The population has been defined as (1,2, . . ., M) and position can be defined as $(x_{i1}, x_{i2}, . . ., x_{in})$ with n dimensions.

2. The initial solution of the monkey is generated in random manner.
3. Climb process
    i) Vector has been generated randomly $\Delta x_i = (\Delta x_{i1}, \Delta x_{i2}, \ldots, \Delta x_{in})$ where Δx_{ij} is set as a, which is denoted as the climb process step length.
    ii) $f'_{ab}(z_a) = \frac{f(z_a + \Delta z_a) - (z_a - \Delta z_a)}{2\Delta z_{ab}}$, $b = 1, 2, \ldots, n$ is denoted as the pseudo-gradient of the fitness function.
    iii) Set $y_b = z_{ab} + a\, sign\left(f'_{ab}(z_a)\right), b = 1, 2, \ldots, ny = (y_1, y_2, \ldots, y_n)$
    iv) Let $z_i \leftarrow y$ when y is viable; otherwise, maintain z_i.
    v) Repeat the preceding steps until there is slight deviation in the fitness function in consecutive steps or an extreme number of steps has been reached, and this is represented by N_c.
4. Watch-jump process
    i) Produce a random number from y_i $(z_{ab} - d, x_{ab} + d), b = 1, 2, \ldots, n$. where d is defined as monkey eyesight.
    ii) Let $z_a \leftarrow y$ if $f(y) > f(z_i)$ and y is viable. Otherwise, redo the above step until a suitable location y is established or a definite quantity of watch periods is attained.
    iii) The climb process is repeated by engaging the initial position y.
5. Somersault process
    i) A real number θ is generated randomly as the somersault interval $[c, d]$.
    ii) Fix $y_b = z_{ab} + \theta(q_b - z_{ab})$, where $q_b = \frac{1}{M}\sum_{i=1}^{M} b = 1, 2, \ldots, n.q = (q_1,$ $q_2, \ldots, q_n)s$, called the somersault pivot and $(q_b - z_{ab})$ is the direction of monkey in the somersault process.
    iii) Set $z_a \leftarrow y$, if $y = (y_1, y_2, \ldots, y_n)s$ feasible. Otherwise, repeat the preceding steps until the suitable solution is found.
6. Termination.

In the Monkey Search algorithm, the population size is defined during initialization. Then the monkey's position is randomly defined, and it is represented using a vector x_i. The climb process is used to change the monkey's position step by step, and this process improves the fitness function of algorithm. After this process, each monkey has reached the top of the mountain and will look at another, higher point, which will be greater than the present location. If the higher position is set up, then the monkey will reach the new position by using its eyesight. Eyesight is one of the parameters that can be explained as the maximum distance watched by monkey. Then a new location is gained, and again the monkey searches for a new position, and this process is called the somersault process. The process can be terminated if the convergence criterion is satisfied.

8.8 Intelligent Water Drop

The Intelligent Water Drop algorithm has been formulated based on the few movements that take place in the river. Velocity and soil are the properties that have been created in this algorithm and the characteristics of these properties can be changed during the course of the algorithm.

In this algorithm, water can be flown from source (initial velocity) to endpoint (zero soil) (Shah-Hosseini, 2009).

Discrete step is followed for the flow of water from the present point to the next point, and the velocity can be increased in non-linear fashion, which is proportional to the inverse of the soil between the present point and the next point. The time interval has been calculated based on the laws of linear motion.

The algorithm works faster whenever the path has less soil when compared with the path that has more soil. It collects the soil throughout the trip and then the soil is removed from the path between the present point and the next point.

The Intelligent Water Drop algorithm requires a mechanism to choose the path to its subsequent location. This mechanism is used to give preference to the path that has low soil when compared to the path that has high soil, by applying uniform random distribution. By using this mechanism, higher preference is given to lower-soil areas.

Most of the Intelligent Water Drop algorithm consists of two parameters, which are static and dynamic. In static conditions all the parameters are constant, and in dynamic conditions, the parameters are reinitialized in every iteration.

8.8.1 Pseudo-Code of Intelligent Water Drop

1. Static parameters are initialized
   a) *Problem formulation through a graphical format*
   b) *Static parameters values are initialized.*
2. Initialization of dynamic parameters soil and velocity
3. Distribution of Intelligent water drop into problem's graph
4. Creation of solution by IWDs end to end with updated soil and velocity
   a) *Local soil is updated in the graph.*
   b) *Soil and velocity are updated on the IWDs.*

5. Local search over each IWD's solution (optional)
6. Updating of global soil
7. Total best solution is updated
8. If the termination condition is not met, go to step 2.

If the properties of convergence appear in the Intelligent Water Drop algorithm, which implies that the number of iterations is large, then the optimal solution can be located.

8.9 Dolphin Echolocation

Griffin introduced the word 'echolocation' to define the ability of the bats to identify the prey and obstacles through echoes emitted by the bats. The best examined echolocation model in nautical animals is dolphins. Sounds created by the dolphins are denoted as clicks. The repetition of these clicks may differ for different species. When the sound reaches the prey, some of the noise returns toward the direction of the dolphin (Lenin, Ravindranath Reddy, & Surya Kalavathi, 2014).

When the echo is received, one more click is created by the dolphin. The time between the click and echo allows the dolphin to evaluate the separation from the object. The changing energy of the signal as it is received on the two sides of the dolphin makes a beeline for assessing the way to the prey. By incessantly transmitting clicks and receiving echoes in this procedure, the dolphin can take off after objects and home in on them. The clicks are directional. From time to time, echolocation happens as a short group of clicks, called the click rate. The click rate increments as the dolphin nears the object. Although bats also utilize echolocation, it might appear differently in relation to dolphins in their sonar plot.

Bats use their sonar plot at little ranges around 3–4 m; however, dolphins can identify their destinations at ranges moving in excess of a hundred meters. A large number of bats pursue bugs that dash rapidly forward and backward, making the pursuit obviously not quite the same as the direct run of a fish sought after by a dolphin. The pace of sound in air is around one-fifth of that of water; in this way, the information transmission rate for the time of sonar transmission of bats is altogether shorter than that of the dolphins.

8.9.1 Dolphin Echolocation Process

Dolphins primarily look all around the hunt space to discover prey. The moment a dolphin approaches the target, the animal limits its request and incrementally fabricates its clicks with a specific end goal to focus on the area. This procedure mirrors dolphin echolocation by constraining its examination according to the separation from the target. Before starting, the observation space should be managed by utilizing the accompanying control in the search space.

8.9.2 Search Space Order

For each variable to be optimized during the methodology, sort the choices of the search space in a tough or downhill order. On the off chance that options assess more than one characteristic, at that point complete requesting is indicated by the most significant one. Utilizing this procedure, for variable j, vector A_j of length LA_j is molded, which contains every plausible option for the jth variable by arranging the vectors subsequent to each other, as the segments of a framework, the matrix alternatives $_{MA + NV}$ are delivered, in which MA is max$(LA_j)_j = 1: NV$, with NV being the quantity of factors. Besides, a bend as per the difference in meeting factor amid the improvement procedure ought to be relegated. Here, the difference in meeting (CF) is considered as

$$PP(Loop_i) = PP_i + (1 - PP_i)\frac{Loop_i^{power} - 1}{(LoopsNumber)^{power} - 1} \tag{8.22}$$

PP is the probability to be predefined, PP_1 is the merging element of the main circle to choose the appropriate response arbitrarily and $Loop_i$ is the quantity of the present loop.

The well-ordered method for dolphin echolocation calculation (DEA) is:
i. Begin the NL areas for a dolphin at your discretion.
ii. Calculate the PP of the loop.
iii. The wellness of every area is figured.
iv. The aggregate wellness, as indicated by dolphin rules, is ascertained.

- For i = 1 to the number of locations
- For j = 1 to the number of variables
- Discover the situation of $L(i,j)$ in jth segment of the alternative's lattice, and name it A. For $k = -R_e$ to R_e

$$AF_{(A+K)j} = \frac{1}{R_e} * (R_e - |k|)Fitness_i + AF_{(A+K)j} \tag{8.23}$$

Where $AF(A + K)j$ is the aggregate fitness, Re is the compelling range. Fitness (I) is the fitness of area I. It ought to be added that for options near edges (where $A + k$ isn't substantial; $A + k < 0$ or $A + k > LAj$), the AF is figured utilizing an intelligent trademark. Keeping in mind the end goal to pass out the alternative equally in the hunt space, a little estimation of $\varepsilon\varepsilon$ is added to all the clusters as $AF = AF + \varepsilon\varepsilon$. Here, $\varepsilon\varepsilon$ ought to be chosen by the strategy as the fitness is characterized. It is superior to be less than the minimum value achieved for the fitness. Locate the best area of this circle and name it "The best location." Discover the options assigned to the factors of the best area, and let their AF be equivalent to zero. Also, it can be defined as:

- For $j = 1$: Number of variables
- For $i = 1$: Number of alternatives
- If $i =$ The best location(j)

$$AF_{ij} = 0 \tag{8.24}$$

v. For variable $j_{(j=\ 1 to NV)}$, figure the probability of picking elective $i_{(i=\ 1 to ALj)}$, as indicated by the accompanying equation:

$$P_{ij} = \frac{AF_{ij}}{\sum_{i=1}^{LAj} AF_{ij}} \tag{8.25}$$

vi. Apportion a likelihood equivalent to PP to all options decided for all factors of the best area and commit the rest of the probability to alternate options as per the accompanying equation:

$$P_{ij} = PP \tag{8.26}$$

Else

$$P_{ij} = (1 - PP)P_{ij} \tag{8.27}$$

vii. Process the subsequent step areas as per the probabilities doled out to every option. Rehash steps ii – vi the same number of times as the Loops Number. This parameter decided for this strategy ought to be a sensible approach.

8.10 Biogeography-Based Optimization

Biogeography-based Optimization (BBO) relies upon the hypothesis of island biogeography. Island biogeography is a field inside biogeography that reviews the factors that impact the species plenitude of isolated normal gatherings (Ammu, Sivakumar, & Rejimoan, 2010). This hypothesis tries to estimate the quantity of species that would exist on a recently made island. An "island" is a domain of sensible conditions incorporated in a spread of unsatisfactory living space. Movement and change are the two principle factors that impact the species' lavishness of islands. Movement incorporates two crucial systems: migration and resettlement. Migration and displacement are impacted by various elements; for example, the separation from an island to the nearest neighbour the size of the island, territory reasonableness record (TRR), and so forth.

TRR incorporates distinctive factors, for example, precipitation, vegetation, environmental condition, etc. These components bolster the nearness of species in a living space. Living spaces that are fitting for the residence of natural species will have high TRR. A natural surrounding with high TRR will be possessed with countless species. So, it will have a high emigration rate and a low immigration rate (since the living space is almost immersed with species). In a like manner, natural surroundings with low TRR will have only a couple of species. This consideration is used as a piece of BBO for completing a movement. In BBO, as in other advancement calculations, at initialization, a broad number of candidate courses of action are created randomly for the issue under consideraton. Related to each arrangement there will be an HSI. Each arrangement delivered is considered to be a living space. Each arrangement is a collection of appropriateness record factors.

Reasonableness list factors (RLF) show the suitability of the natural surroundings as a living space. High HSI living space is, for all intents and purposes, proportionate to a decent arrangement, and low HSI environment is like the poor arrangement. Through relocation, a high HSI arrangement imparts a great deal of highlights to poor arrangements, and poor arrangements can recognize an impressive measure of highlights from great arrangements. The connection between species check, movement rate, and displacement rate is shown in Figure 8.2 on the next page, where I allude to the greatest migration rate, E is the most extreme resettlement rate, S_0 is the harmonious number of species, and S_{max} is the most extreme species tally. The decision to modify each arrangement is taken in perspective of the movement rate of the arrangement. Transformation is the sudden change made to the TRR of any living space because of a specific occasion. Transformation expands the variety of the populace. Every hopeful solution is related with a transformation likelihood, as characterized by condition 8.28.

$$M(s) = M_{max}(1 - P_s)/P_{max} \qquad (8.28)$$

Where M_{max} is a client-characterized parameter, P_s is the species tally of the living space; P_{max} is the most extreme species tally. Transformation is done in view of the change likelihood of every natural surrounding by

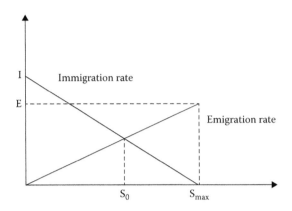

FIGURE 8.2
Relationship between species count, immigration rate, and emigration rate.

supplanting an SIV from the living space with another randomly produced RLF.

BBO calculation can be casually characterized as follows:

1. Characterize the movement likelihood and transformation likelihood.
2. Population initialization.
3. Ascertain the migration rate and resettlement rate of every competitor in the populace.
4. Select the island to be altered in light of the migration rate.
5. Using roulette wheel selection on the emigration rate, select the island from which the RLF is to emigrate.
6. Randomly select an RLF from the chosen island to emigrate.
7. Perform change in view of the transformation likelihood of every island.
8. Figure the wellness of every individual island, on the off chance that the wellness standard isn't fulfilled

Go to step 3.
Immigration rate R_i can be defined as

$$R_i = I(1 - F(s)/n) \tag{8.29}$$

Emigration rate R_e can be defined as

$$R_e = E(F(s)/n) \tag{8.30}$$

Where I is the immigration rate, E is the emigration rate, $F(s)$ is the wellness rank of arrangement s, and n is the quantity of applicant arrangements in the populace.

8.11 Paddy Field Algorithm

The PFA is another organically inspired algorithm, which simulates the development procedure of paddy field (Gandomi, Yang, Talatahari, & Deb, 2012). In an uneven paddy field, generally the seeds are sown and the factors affecting the growth of the seeds are fertile soil and soil moisture. The above factors certainly have effects on the development of the seed. The growth of the plants depends on the placement of the seed, or where it falls. The seeds falling near fertile soil and abundant soil moisture will have the best growth of plants, which deliver the most seeds. At the point when the seeds move toward becoming plants, their reproduction depends on the pollination effects. The pollen, which is regularly conveyed by the breeze, identifies with the population density. The places where there is a high density of the plants are expected to generate more pollen and

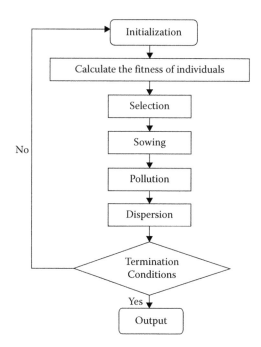

FIGURE 8.3
The structure of PFA.

produce more seeds. The plants that create the most seeds are viewed as the optimal solution of the optimization problem. PFA simulates the progressive activity of a paddy field, producing applicant solutions that grow toward the best solution as shown in Figure 8.3 on the previous page.

PFA does not encourage crossover among individuals, and it is very similar to GA. The implementation of this algorithm through programming is simple. It operates on a generative principle reliant on closeness to the universal solution, and density like that of a plant population. The fitness function is $a = f(b)$, where $b = [b_1, b_2, \cdots\cdots, b_n]$. The vector $b = b^i$ is called as one seed plant and a represents its fitness value. Each dimension of a seed must be within b_j ε $[x, y]$. The standard PFA, which simulates the growth process, has five parts, namely initialization, assortment, propagating, pollination, and disseminating. Bad seeds are eliminated after a number of iterations, whereas the good seeds are deposited to produce more good seeds. During this iteration cycle, aspirant solutions move toward the better solution.

8.12 Weightless Swarm Algorithm

The working process of the Weightless Swarm algorithm (WSA) is similar to that of particle swarm optimization (Ting, Man, Guan, Nayel, & Wan, 2012). Without the weight in inertia, the multiple line equation is simplified to one line and the updated equation is shown as

$$X_{i,d}^{t+1} = X_{i,d}^t + r_1 z_1 (Gbest_{i,d} - X_{i,d}^t) + r_2 z_2 (Pbest_{i,d} - X_{i,d}^t) \qquad (8.31)$$

Where $X_{i,d}$ is the location of dth dimension of the ith particle. $Pbest$ is the finest location of a particle during exploration. $Gbest$ is the finest result in exploration history. r_1 and r_2 have been denoted as random independent variables between $[0, 1]$. z_1 and z_2 are acceleration coefficients set toward 1.7. The constriction factor $N = 0.729$ is initialized based on $|z_1 + z_2| \le 4.1$. Assume $z_1 = z_2 = 2.05$ and $N = 0.729$, coefficients will be $0.729 \times 2.05 = 1.49445$. Therefore, new solutions represented on the basis of the dynamic systems principle for investigation of a particle path have been passed out with variable parameter ($e = 0.6$ and $z_1 = z_2 = 1.7$), which displayed better performance. WSA approved these constraint settings without an inertia weight parameter. Assign $z_1 = z_2 = 1.7$, outcomes are somewhat better when equated to 1.5. By default, the value for z_1 and z_2 for PSO in evolutionary algorithm is also 1.7. Based on the results obtained through static arithmetical problems, an equation (8.33) can be found, and it has been further simplified as

$$X_{i,d}^{t+1} = X_{i,d}^t + r_1 z_1 (Gbest_{i,d} - X_{i,d}^t) \qquad (8.32)$$

Equation (8.34) works well when swap has been done through the *Pbest* and *Gbest* updated values. So, in WSA various parameters are lost, such as inertia and velocity, which are mostly used in PSO. Without velocity, one can reject the parameters, which are V_{max} and V_{min}. The acceleration coefficient can also be removed in WSA. Thus, this algorithm is simple to implement compared to PSO. The major advantage of this algorithm is the reduced complexity, and the user has to tune z_1 for better performance, lower computational cost. The un-discloser of WSA is extremely simple.

The implementation of WSA can be done through the existing PSO algorithm based on the following steps:

i. Inertia weight is set to 0 during initialization.
ii. Remove the *Pbest* parameter by setting z_1 in Equation (8.32) to zero.
iii. *Pbest* is updated during the swap process. For *Gbest*, swapping is not required because it will take one of the *Pbest* values as *Gbest*.

Thus, the WSA is implemented and the performance of the algorithm is improved without the inertia weight parameter.

8.13 Eagle Strategy

The Eagle Strategy algorithm is formulated as a two-stage process. The two stages are a crude global search and an intensive local search. During the first stage of the algorithm, it explores the global search space with the help of Levy flight random walks, and if the solution is found, then the second stage process is employed by using intensive local search methods, such as the downhill simplex algorithm and hill climbing algorithm (Kong, Chen, Xie, & Wu, 2012). Again, the process is repeated until the eagle finds the promising region. The benefits of a two-stage process include: a tradeoff has been achieved with the help of global search and a fast local search. Further, one can use any algorithm throughout the search. In metaheuristics, the so-called Lévy distribution is defined as the distribution of the addition of M identical and independent distributed arbitrary parameters whose characteristic function can be written using the Fourier transform:

$$F_M(Z) = e^{[-M|Z|^\beta]} \tag{8.33}$$

Lévy distribution denoted by $L(S)$ with an index α and to get the actual distribution, the parameter $L(S)$ the inverse of the solution is not straightforward because of the integral factor. For most of the applications, one can assign $\beta = 1$ for easiness. For the special cases, assign $\alpha = 1$ and $\alpha = 2$.

When $\alpha = 1$, the above integral turns out to be the Cauchy distribution. When $\alpha = 2$, it turns out to be the normal distribution. Therefore, Lévy flights turn into the standard Brownian motion.

$$L(S) = \frac{1}{\Pi} \int_0^\infty \cos(\tau s) e^{-\beta \tau^\alpha} d\tau, (0 < \alpha \le 2) \tag{8.34}$$

In the second stage, one can apply differential evolution in place of the intensive local search, in contrast to the gradient-based method like hill climbing. It is well known that differential evolution is fundamentally a global search algorithm, and it can simply be adjusted to an efficient local search by restricting the new solutions locally near the best promising region. The advantage of using a differential evolution algorithm is that it is derivative-free. The fundamental steps of eagle strategy along with different evolution algorithms are presented below. Based on the discussion, one can look at every iteration loop — first a global search is carried out, then the local search follows.

Pseudo-Code
        Set the objective function
        Random initial guess has been set to $X_t=0$
        While (end criterion)
        Randomizing global exploration using levy flights
        Objectives are evaluated to find a better solution using differential
            evaluation
        If an optimum solution is found
        Current best has been updated
        End
        t=t+1 has been updated
        End
Post-process the result and visualization.

8.14 Conclusion

Miscellaneous swarm intelligence techniques based on the behaviour models of termites, cockroaches, bumblebees, spiders, cats, monkeys, dolphins, and eagles are discussed with respect to the applications in the field of computer science for global unconstrained optimization problems. Miscellaneous swarm intelligence techniques have been highlighted based on their working methodology, algorithms, and mathematical modeling with a brief explanation. The chapter is focused in such a way as to address the algorithms for solving the problems in the industrial applications. It should be noted that this chapter does not cover all the details about the miscellaneous swarm intelligence techniques; however, the basic principles of the algorithm have been discussed precisely. The main objective of this chapter is to deliver an overall summary to the readers who are interested in miscellaneous swarm intelligence techniques.

References

Ammu, P. K., Sivakumar, K. C., & Rejimoan, R. (2010). Biography-based optimization-A survey. *International Journal of Electronics and Computer Science Engineering, 2*(1), 154–160).

Bahrami, M., Bozorg-Haddad, O., & Chu, X. (2018). Cat Swarm Optimization (CSO) algorithm, advanced optimization by nature-inspired algorithms. *Studies in Computational Intelligence* (Vol. 720, pp. 9–18). DOI: 10.1007/978-981-10-5221-7_2.

Bonabeau, E., Dorigo, E., & Theraulaz, G. (1999). *Swarm Intelligence: From Natural to Artificial Systems*. Oxford, UK: Oxford University Press, pp. 1–278.

Gandomi, A. H., Yang, X.-S., Talatahari, S., & Deb, S. (2012). Coupled eagle strategy and differential evolution for unconstrained and constrained global optimization. *Computers and Mathematics with Application, 63*, 191–200.

Hooker, J. N. (1994). Logic based methods for optimization, principles and practice of constrained programming In *Lecture Notes in Computer Science* (Vol. 874, pp. 336–349). Springer Publishing.

Kong, X., Chen, Y.-L., Xie, W., & Wu, X. (2012). A novel paddy field algorithm based on pattern search method. In *IEEE Proceedings of the International conference on information and automation*, pp. 686–690.

Kwiecien, J., & Pasieka, M. (2017). Cockroach swarm optimization algorithm for travel planning. *Entropy, 19*, 1–15.

Lenin, K., Ravindranath Reddy, B., & Surya Kalavathi, M. (2014). Dolphin echolocation algorithm for solving optimal reactive power dispatch problem. *International Journal of Computer, 12*(1), 1–15.

Marichelvam, M. K., Tosun, O., & Geetha, M. (2017). Hybrid monkey search algorithm for flow shop scheduling problem under *Make Span* and total fow time. *Applied Soft Computing, 55*, 82–92.

Marinakis, Y., Marinaki, M., & Matsatsinis, N. (2010). A Bumble bees mating optimization algorithm for global unconstrained optimization problems. *NICSO, 284*, 305–318.

Parpinelli, R. S., & Lopes, H. S. (2011). New inspirations in swarm intelligence: A survey. *International Journal of Bio-Inspired Computation, 3*, 1–16.

Shah-Hosseini, H. (2009). The intelligent water drops algorithm: A natural inspired swarm-based optimization algorithm. *International Journal of Bio-Inspired Computing, 1*(1–2),71-79.

Ting, T. O., Man, K. L., Guan, S. U., Nayel, M., Wan, K. (2012). Weightless swarm algorithm (WSA) for dynamic optimization problems. In J. J. Park, A. Zomaya, S. S. Yeo, S. Sahni (Eds.), *Network and Parallel Computing. NPC 2012. Lecture Notes in Computer Science*, vol 7513. Berlin, Heidelberg: Springer.

Yu, J. J. Q., & Li, V. O. K. (2015). A social spider algorithm for global optimization. *Applied Soft Computing, 30,* 614–627.

Zungeru, A. M., Ang, L.-M., & Seng, K. P. (2012). Termite-hill: Performance optimized swarm intelligence-based routing algorithm for wireless sensor networks. *Journal of Network and Computer Applications, 35,* 1901–1917.

9

Swarm Intelligence Techniques for Optimizing Problems

K. Vikram and Sarat Kumar Sahoo

CONTENTS

9.1 Introduction

Modern research is attaining optimistic solutions for many complex engineering problems, on the basis of the social insect metaphor. These solutions, inspired from biological systems, emphasize the broadness of the communication (direct or indirect) among the different entities. The applications aim for the artificial intelligence based on swarm intelligence by selecting some set of the entities chosen. The outcomes obtained are providing the greater flexibility and robustness for many engineering applications such as communication networks, military applications, and robotics.

Swarm intelligence is based on the structure of systematic activities going on inside insect colonies. The insect may be an ant, bee, or termite. It is very interesting to see every individual work within its own specifications without a supervisor's order. But there arises a question: Who governs? Who will give orders, do an estimation of the future, develop strategies, and maintain equilibrium? The above is a puzzling question. Each individual in the swarm has its own agenda. Even though all the individuals are working on their own, they do it in an organized fashion. The integration of all the individuals in the colony is very interesting.

The importance of this chapter is understanding how to design, decentralized and flexible algorithms for artificial intelligence, communication networks, robotics, data mining, big data analytics, and the Internet of Things (IoT) on the basis of swarm intelligence. The modeling of the above-mentioned artificial systems based on the collective behaviour in insects is the first step in understanding the essence of this chapter.

9.2 Swarm Intelligence for Communication Networks

9.2.1 Introduction

The management of communication networks is becoming much tougher because of the vast network size, swiftly changing topology, and complexity. There is a necessity for integration of the many heterogeneous components in the network, and this task is indeed a challenge. Integration and interoperability are extremely difficult for networks that traditionally follow the integrated methods for network control. This problem is common for both the virtual circuit and packet switched networks. The reliability of communication network depends on the routing algorithms. The modern network routing techniques are based on the performance metrics like average throughput, and delay, and should ensure the quality of service (QoS) parameters (Kassabalidis, El-Sharkawi, Marks, Arabshahi, & Gray, 2001).

Swarm intelligence gives rise to complicated and often intelligent performance through the complex communication of thousands of self-governing swarm members. The main ideology behind these interactions is stigmergy (interaction via the environment). The stigmergy is an intelligent task-based behaviour that alters the environment so as to encourage further parallel actions by the swarm members. For example, when constructing the nests, the termites lay a sand grain at random locations; indeed they are capable of systematic designing (Gui, Ma, Wang, & Wilkins, 2016).

This swarm intelligence is very useful for emergent behaviour situations that arise in the communication networks. Emergent behaviour occurs because of the simple collaboration of independent nodes, with simple primitives that result in complex behaviour. Swarm intelligence provides many benefits due to the employment of mobile nodes and stigmergy. Scalability, adaption, error tolerance, rapidity, modularity, independence, and parallelism are the advantages of swarm intelligence for communication networks (Çelik, Zengin, & Tuncel, 2010).

Swarm intelligence in communications can be explained on the basis of the following types:

a. **Ant colony optimization:** A set of optimization algorithms that demonstrates behaviour based on the activities of an ant colony. This model is very suitable for finding the paths to goals. The simulated ants in the colony serve as simulation agents to trace optimal solutions by relocating over a parameter space searching for all the possible solutions. Real ants put down pheromones guiding each other to sources of food, while discovering their environment. The replicated ants correspondingly record their locations and the quality of their discoveries, so that in future simulation repetitions more ants find better clarifications (Ma, Sun, & Chen, 2017; Zheng, Zecchin, Newman, Maier, & Dandy, 2017).

b. **Particle swarm optimization (PSO):** A global optimization methodology for handling the problems of a specific pattern, it aims to obtain the best solution among many available solutions. The advantage of this model is that it is more useful when there are a large number of members that constitute the particle swarm, making the system impressively strong to resist the problem of local minima. In the communication networks, PSO is more useful for finding the best routing that helps in the energy efficient operations (Javan, Mokari, Alavi, & Rahmati, 2017).

c. **Stochastic diffusion search (SDS):** A proxy-based probabilistic universal search and optimization technique, it is suited to problems where the objective function can decompose into many independent partial functions (Kassabalidis et al., 2001). An SDS is defined based on distributed computation, where the actions of simple computational units, or agents, are inherently probabilistic. Agents supportively construct the solution by execution of autonomous searches, surveyed by the distribution of data through the population. Positive feedback encourages better results by assigning to them more agents for their investigation. Limited resources provoke strong opposition from which the major population of agents equivalent to the best-fit solution rapidly emerges.

9.2.2 Swarm Intelligence Routing in Communication Networks

Swarm intelligence in the communication network is more useful for finding the optimal routing based on the (Bastos-Filho, Schuler, Oliveira & Vitorino, 2008):

a. Ant-based control (ABC)
b. AntNet

a. Routing based on ABC (Kassabalidis et al., 2001; Tatomir, Rothkrantz, & Suson, 2009)

-- The ABC algorithm is characterized by evolution and reveals the robustness in various network conditions.
-- It is carried out by means of agents, known as ants, who apply a sediment of virtual pheromone on the tracks they follow. If we model this in the mathematics, then these phenomena affect the routing tables of the affected nodes.
-- The probabilities change based on the ant visits; hence, the routing tables must be updated.
-- A lifetime of the ant at the time of the visit (T): $T = \sum_i D_i$

    o Delay (D_i):

$$D_i = c.e^{-d.S} \tag{1}$$

-- The step size for the visited node (δ_r) : $\delta_r = \dfrac{a}{T} + b$
-- The routing table is then revised as follows

    o

$$r_{i-1,s}^i \, (t+1) = \frac{r_{i-1,s}^i \, (t+1) + \delta r}{1 + \delta r} \tag{2}$$

    o

$$r_{n,s}^i \, (t+1) = \frac{r_{n,s}^i \, (t) + \delta r}{1 + \delta r} n \neq i - 1 \tag{3}$$

-- The ant node utilizes and updates the routing table at the same time.
-- For example, if node G is the source node and node H is the destination, then the ant will inform the row for G and informs node H to find the next route

    o Repeat for updating rules $\sum_n r_{n,s}^i = 1$

-- Cooperation of an investigation factor h

    o With probability $(1 - h)m$ the ants are promoted with an overall uniform distribution

○ With probability *h*m the ants are promoted based on the routing table probabilities

Routing is defined by means of extremely complicated relations for forwarding and backward based system exploration agents "ants." The concept of the subdivision of the agents is to allow the backward ants to consume the useful information grouped by the frontward ants on their tour from initial source to the destination. The probability values of the routing table must sum to "one" for each row of the communication network. The obtained probabilities will be useful in two ways, the exploration agents (ants) of the network uses the obtained probabilities to decide the next route to reach the destination. The casually selected candidates on the basis of the routing table probabilities for a precise endpoint deterministically find the path with the maximum probability for the next route.

b. Routing based on the AntNet (Kassabalidis et al., 2001; Karia & Godbole, 2013). The categorization of activities in the swarm is simple

   i. Each swarm member in the network introduces the forward ant's concept to all destinations points in the area considered at regular time intervals.

   ii. The swarm member (ant) discovers a route to the destination in a random order on the basis of the current existing routing tables.

   iii. The forward ant generates a stack knowledge on routing, mandating in trip periods for every swarm member as the node is going to reach the destination.

   iv. When the swarm member reaches the destination, the backward ant receives the stack knowledge.

   v. Then the backward ant reports the stack knowledge entries and tracks the path in reverse order.

   vi. The node tables of every visited node are updated with recent data based on the trip times.

Each swarm member in the network has a knowledge on the routing table; each swarm member also possesses a stack table with track records of the mean and variance of the route to every destination. Figure 9.1 shows the categorization of different swarm intelligence-based routing protocols (Çelik et al., 2010).

9.3 Swarm Intelligence in Robotics

The swarm robotics (SR) demonstrates great ability in several aspects, which are motivated by nature. SR is an amalgamation of the swarm intelligence and

		M. Dorigo et al. (1998)
		Y. Zhang et al. (2004)
	Ant Colony	L. Juan et al. (2007)
	Optimization based	A. M. Zungeru et al. (2012)
		P.V. Krishna et al. (2012)
Swarm		B. Singh et al. (2017)
intelligence-based		S. Ramesh et al. (2013)
routing protocols		J. Loganathan et al. (2015)
for WSN	Bee Colony based	S. Kaur et al. (2017)
		Xiu Zhang et al. (2017)
		M. Phone et al. (2007)
	Slime Bold based	Ke Li et al. (2011)
		M. Zhang et al. (2014)

FIGURE 9.1
The categorization of different swarm intelligence based routing protocols (Çelik, Zengin and Tuncel, 2010).

robotics. The swarm robotic algorithms are based on the cooperative control mechanisms for clustering, circumnavigating, and searching applications (Karia & Godbole, 2013). Swarm intelligence technology is inspired by the biological swarms and implements the system as an assembly of the swarm instead of treating it as an unconsidered individual. A swarm intelligence system includes set of simple objects independently controlled by a simple set of instructions and indigenous relations. These entities in the swarm are not essentially unwise but are very modest compared to the global intelligent systems. There are some aspects where an individual starts acting very wisely when several individuals begin to cooperate (Suárez, Iglesias, & Gálvez, 2018).

9.3.1 The Swarm Robotics – Definition

SR is a modern approach for the synchronization of multiple-robot systems with many simple physical robots. The anticipated behaviour develops from the communication among the robots and collaboration of robots within the location. This method appeared in the area known as artificial swarm intelligence, inspired by biological learning based on swarm behaviour in nature. The present research on the SR is mainly focusing on the design paradigms of comparatively very simple robots, their controlling architectures, and physical design. In a general swarm, individuals are modest, insignificant, and available at low cost, to be used with the intention of taking advantage of a large population. The most important aspect of this approach is to utilize the effective communication among the agents of the swarm that is available locally and ensures the system's robustness and accessibility. The simple rules for the individual levels can generate a large collection of complicated behaviour at the swarm level. The rules are framed by regulating the individuals and are extracted from the cooperative behaviour of swarm (Bayindir, 2016).

The important applications of swarm robotics consist of jobs that demand a reduction in the size of the robots and distributed sensing jobs in micro-machinery. Swarm robotics is more suitable for applications like mining activities and agriculture foraging jobs. Swarm robotics is also engaged in work that requires a large place and time cost, and is hazardous to a human being. They work well for target searching and military applications (Senanayake, Senthooran, Barca, Chung, Kamruzzaman, & Murshed, 2016).

9.3.2 Modeling Swarm Robotics

In the swarm robotics model information exchange between every individual in the swarm is based on sensing, communication, and motioning. In the swarm, the exchange of information happens between robots, and this results in swarm-level coordination.

a. Information exchange module:

The controlling of robots in the swarm can be easy because of their information exchange with one another. The important functions of the individual in the swarm are limited sensing and limited communications. The strategies can be different, based on the different applications in the swarm. The information extracted from the environment will be studied and processed, and the robots acknowledge the messages back to the environment; their pheromones act as an environment here (Dorigo, Bonabeau, & Theraulaz, 2000).

b. Direct communication:

The direct communication of the network is of two types: Ad-hoc network and broadcast. There exists many wireless technologies and protocols, but specifically for the SR, the topologies still remain undiscovered. The existing network protocols for computer networking are designed for data communication and information exchanges between the nodes. The communication network in the SR should consider local sensing and signaling abilities and should also consider boosting the cooperative behaviours of the individual swarm members (Hawick, James, Story, & Shepherd, 2002).

c. Communication through the environment:

The swarm robots can communicate with each other through the existing environment. The natural environment acts as the communication medium between the robots for the interaction. The robots in the swarm do not communicate directly with each other. But they stimulate their neighbouring swarm members, which can sense the trace of the preceding robot. In this way, the sequence of the actions can be systematized, and this leads to spontaneous swarm-level activities. The SR is inspired by the natural swarm members like ants or bees that communication by means of pheromones.

Swarm Robots (SR) cannot adapt the phermone-based natural phenomena but can adapt the similar process for the communication purpose based on the adavanced intelligent automation process. Ranjbar-Sahraei, Weiss, and Nakisaee (2012) have proposed a coverage approach with the help of natural swarm members that do not have direct communication and are communicating naturally based on the swarm intelligence principles. Payton, Estkowski, and Howard, (2005) have implemented an SR system that follows a type of biologically stimulated signaling, called "virtual pheromone," for distributed computing. Grushin and Reggia (2006) have resolved a problem of self-association of prespecified 3D architecture from the blocks of different proportions with an SR system, using stigmergy.

d. Sensing:
The individuals in the swarm can sense the difference between the robot or the other environment entities. Each swarm consists of the onboard sensors that are responsible for sensing the objects or targets. The onboard sensing must perform obstacle avoidance, target exploring, and the flocking. The main purpose of the sensing is to integrate the swarm members with good sensing ability. Cortes, Martínez, Karatas, and Bullo (2004) have discovered the way of controlling and coordinating a group of independent vehicles. The vehicles are considered to be the swarm members with sensors, in a simple, spread out, and asynchronous way. The effective communication requires more sophisticated hardware and systematic synchronization, with the best bandwidth and energy. Also, the system must allow expansion if the swarm population grows. SR working with a cooperative model should work with simple communication and should utilize proficient sensing for best solutions.

e. Cooperation schemes between robots:
The cooperation schemes play a very important role in the SR model. In SR, cooperation takes place at two levels: at every individual level and at the swarm level on the whole. The former is essential for the robot's actions and necessary for the coordination of the inputs. The inputs for every robot are supplied from the environment with respect to response, absorbing, and adjusting behaviours. The latter cooperation scheme is a combination of former cooperation efforts; this results in the tougher collection of the tasks such as group, disperse, or formation. Several sub-problems have been proposed for cooperation between robots, which are described in detail in this section.

f. Local coordination:
In the SR there are no global coordinating systems. Therefore, each individual in the swarm must maintain the local coordinating system efficiently. The local coordinating system should work efficiently for identification, distinguishing, and locating the nearby robots. For maintaining this coordination of the sensors, onboard sensing should work efficiently (Borenstein, Everett, Feng, & Wehe, 1997).

g. Physical connections:
In certain cases, like those where there are large gaps between the swarm members, and cases like cooperative transportation, the physical connections are very important for the single robot. The are many existing methods for physical connection in the case of SR. These are using infrared rays for localizing, sensors, and actuators for overcoming the gaps.

h. Self-organization:
In the case of the SR, a self-organization scheme is very useful for building global communication through the local interactions of the swarm members. The swarm-level structure emerges from the individual level.

Table 9.1 below shows some examples of the robotics entity projects based on the swarm intelligence.

9.4 Swarm Intelligence in Data Mining

The data mining is very important for modern intelligent systems and for their operation. Finding very useful data patterns from a huge amount of data available is the main purpose of data mining. For this, Fayyad, Piatetsky-Shapiro, and Smyth (1996) proposed many useful algorithms for data pattern extraction. For the logical extraction of the useful data, there is a need for

TABLE 9.1

Swarm robotics entity projects

Name of the Project	Developed by	Application
Project SI	Embedded Lab of Shanghai Jiaotong University	Designed a project based on SR, called eMouse. Able to test all the swarm-inspired algorithms, communication protocols
Sambots	Hongxing Wei et al., Beihang University	Demonstrated the self-assembly of the individual robots in a swarm
Swarm-bots Project	Future & Emerging technologies, European Commission	Self-organizing and self-assembling of the robots.
Pheromone robotics project	David W. Payton et al., DARPA Software for Distributed Robotics	Scalable approach to achieving the swarm-level, behaviours can find any applications
I-swarm project	H. Wörn et al.	Combines micro-robotics, and distributed and adaptive systems as well as self-organizing biological swarm systems

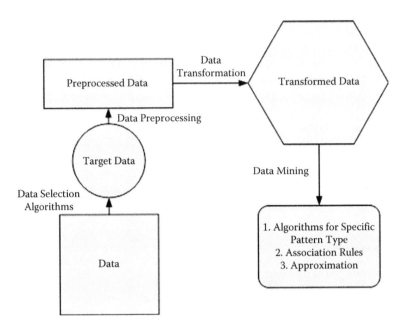

FIGURE 9.2
The systematic procedure involved in the knowledge discovery process.

appropriate specific patterns based on data selection methods, data deletion knowledge, pre-processing the data, and finally, absolute interpretation of the data, which is very useful for ensuring the correct and most useful information is extracted (Bonidia, Brancher, & Busto, 2018). Figure 9.2. shows the systematic procedure involved in the knowledge discovery process.

9.4.1 Steps Involved in Knowledge Discovery

Based on the above diagram (Wang, Li, Zhou, Wang, & Nedjah, 2016), the elementary procedure involved in data mining process is presented as below:

1. The development of an algorithm and interpretation of the application data domain are significant in the knowledge discovery process.
2. Based on the data selection algorithms, target data is set.
3. The data received need to be filtered or pre-processed to make it efficient and useful.
4. Huge data demands huge memory. Hence, one needs algorithms for data size reduction and projection.
5. Managing the goals of the knowledge discovery procedure for a specific data mining method by prediction and description.

To achieve the knowledge discovery using data mining, the following definitions are necessary to understand:

- Clustering will make swarm members understand different categories of the data that they are going to deal with.
- The Summation is an individual capacity of a swarm member to understand the data and its subsets.
- Dependency modeling is discovering a model to illustrate significant communication.
- Regression is about learning a function that defines the sets of the data item.
- Classification is about learning a function that differentiates a particular data item into one of several predefined classes.
- Variation and deviation discovery are all about understanding the most important variations in the data.

9.4.1.1 Swarm Intelligence in Data Mining

Data mining and particle swarm optimization (PSO) have may significant properties in common. In spite of difficulties, many people have used the PSO for data mining. The PSO methods for image processing and pattern recognition were used previously. A PSO based new clustering is applied for image segmentation, in modern image processing, colour image quantization, and mixing problems (Urso, Fiannaca, La Rosa, Ravì, & Rizzo, 2018).

Pictorial data mining deals with an illustration of data and knowledge that comprise the result of optimization problems. For resolving issues like the ones above, hybrid techniques have been presented, based on PSO in combination with traditional optimization methods. Testing and understanding the pictorial data sets in relation to Alzheimer's disease demands a high pictorial resolution for superior illustration, and it can be achieved by merging PSO with conventional optimization approaches.

In order to approximate the worth of PSO for data mining, an experimental assessment is presented using three alternatives combining PSO with another genetic algorithm. First, PSO-data mining for the organization of the responsibilities is utilized. Such responsibilities are measured for core instruments for decision support schemes in an extensive area, extending from the commerce, industry, scientific, and military areas. The information sources utilized here for experimental analysis are commonly used and studied as a de facto criterion for rule finding processes based on consistency ranking. The outcomes obtained in these areas seem to specify that PSO algorithms are inexpensive when used with other evolutionary methods and can be certainly applied to more challenging problem domains.

The Ant K-means (AK) clustering technique alter K-means for detecting the entities in a cluster. The probability of pheromones needs to be updated. The rule of revising the information on pheromone is based on the total cluster variance. A novel clustering technique known as ant colony optimization in combination with a special-favor algorithm functions well when compared to the fast-self-arranging map-based K-means approach and inherent K-means algorithm (Angiulli, 2018). Another algorithm, called a bottom-up method was proposed for keeping the system informed, for getting better results based on time series segmentation. Investigation results show the bottom-up method is working better in terms of time series segmentation when compared to the ant colony optimization algorithm.

9.4.1.2 Data Mining and Ant Colony Optimization

Ant colony-centered clustering algorithms were initially presented by Deneubourg, Goss, Franks, Franks, A. S., Detrain, and Chretien (1991) by imitating dissimilar types of logically happening embryonic phenomena. The ants collect items to form masses (collecting of dead cadavers) detected in the types *Lasius niger* and *Pheidole pallidula*. The basic means motivating this type of combination phenomenon is a temptation arbitrated by ant workers in small clusters, among dead objects, toward storing substances, which grows with the interest of workers in depositing more items. This is very optimistic phenomena and an auto-catalytic response that proceeds to the creation of larger clusters.

Handl, Knowles, and Dorigo (2006) projected a novel method constructed on ant-based clustering. It integrates adaptive, diverse, and time-dependent ants performing an activity. This improves the spatial implanting formed by the algorithm and avoids ambiguity in partitioning. Empirical results validate the capability of ant-based clustering. This helps in the identification of the number of clusters essential for the data collection, and delivers high-worth solutions.

Web usage requests utilize the mining-based efforts to separate the valuable information from the secondary facts and can be attained from the influences of the workers with the Web. Web-based mining has been a serious difficulty for Website administration-based operations when making adaptive Web sites, for commercial and assistance services. Abraham and Ramos (2003) anticipated an ant-clustering algorithm to understand Web usage patterns and a direct genetic program design style to investigate the visitor trends. Experimental outcomes demonstrate that ant colony grouping performs well when associated to a self-organizing map.

9.5 Swarm Intelligence and Big Data

The modern-day research on the big data is very interesting. The data emerging from IoT, wireless body area network, smart grid monitoring, and social networking is so great that the huge volumes of the data cannot be processed easily. The size of the data obtained is huge and is very much greater than the processing capability of a database. The four important components the data deal with are capture, memory, management, and assessment. The handling of large data analysis in a limited time is becoming a great problem. The data is expected to be received from the different resources in some large data sets. In general, more than one objective has to be fulfilled in an instance. With the introduction of swarm intelligence into the big data, many problems can be modeled easily. Data analytics using swarm intelligence plays a key role in managing the large quantity of the data and managing high-dimensional data, it can manage high-dynamic data and can handle multi-objective optimization (Cheng, Shi, Qin, & Bai, 2013). "The useful knowledge is to be extracted from the large data" is the main motto of big data analytics. Extraction of the data can also be called "mining." The research on the big data analytics must be concentrated on the large volumes, the variability of the different sources, and the velocity, i.e., increasing speed. The algorithms aimed at the big data should ensure huge-volume data sets and need to provide the optimal solution for handling them (Wei & Wang, 2017). The process of converting the raw data obtained from the different resources into useful information is called "knowledge discovery in databases (KDD)." The KDD is subdivided into data mining, which endeavors to discover necessary information in the large databases. Data mining consists of subfields like classification analysis, alliance analysis, and clustering analysis (Cheng, Chen, Sun, Zhang & Tao, 2017). Figure 9.3. shows the process of KDD.

Data mining for big data using swarm intelligence is a very new subject in the field of the computational intelligence. It deals with the collective intelligence for the enhancement of a simple individual or a group. Swarm intelligence may be based on PSO or ACO and is utilized for solving the data mining issues with multiple objectives or a single objective. The particle swarm works with two important ideologies; the first one is self-cognitive and the other is social learning. The particle swarm was helpful for the data clustering, document clustering, high-dimensional data and the web data mining (Cheng, Zhang, & Qin, 2016).

9.5.1 Swarm Intelligence in Big Data Analytics

The modern research on the information processing deals with big data analytics. The associations among big data analytics and swarm intelligence can be established and shown in Table 9.2.

In the swarm intelligence algorithms, many solutions are found at the same time. This is very important for big data analytics for handling the huge amounts of the data instantaneously. The volume of huge data received is getting the attention of many researchers for identifying better

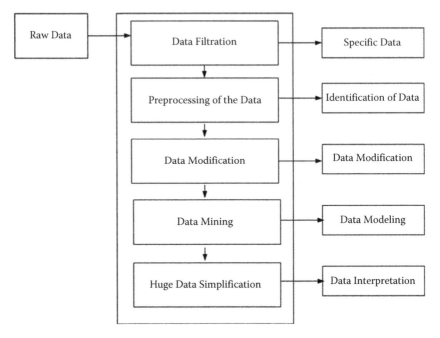

FIGURE 9.3
The process of KDD.

ideas. The various objectives involved will increase the number of problems faced during the data simplification process (Ramírez-Gallego, Fernández, García, Chen & Herrera, 2018). There are three important kinds of
problems we come across with big data, which are:

1. **Large-Scale Optimization**
   Big data is in need of faster mining techniques for processing the
   large-scale data sets that are encountered from the instant generation
   of the data from various resources. The responses of data mining
   should be available in a specific time. For a given problem in big
   data, the solution can be obtained in a permissible time. The solution
   may have a single or several optimal solutions, and the algorithm may
   also give many local optimum solutions for the existing problems. The
   difficulty of the problem will rise with the growing number of variables and aims. The solution limit of a problem often increases with
   the increase of the problem dimension. It requires more efficient
   searching functions to explore the promising variables within the
   specified time. The evolutionary computations based on the swarm
   intelligence are built on the group of solutions. The features of this
   problem may certainly change with respect to the scale. Problems

TABLE 9.2

The Rough Association between Swarm Intelligence and Big Data Analytics.

Big Data Analytics	Swarm Intelligence
Variety	High dimension/large scale
Volume	
Veracity	Uncertainty/surrogates/noise
Value	Objective/ fitness
Velocity	Dynamic environment

become more complex when the dimension rises. The important third direction is based on the good solutions for the identified problems. Swarm intelligence considers each update from every dimension and evaluates them on the whole dimensions' strategy. In the swarm intelligence, the individual result is updated feature by feature. The fitness objective value must be planned for the whole resolution. The solution upgrade depends on the combination of several vectors, i.e., the current value, the difference between current value and previous best value, the differential between the current value and a neighbour's best value, or the difference between two random solutions, etc.

2. **Handling High-Dimensional Data**
   The modern algorithms for handling the data are designed for the present conditions, and they're forecasting on the future conditions. The present algorithms may not be fit future conditions. The nearest neighbour approach is very effective in data categorization. The high-dimensional data is very problematic for solving comparison search problem due to the high computational difficulty problem. The data mining problem can be solved by transforming it into an optimization problem. Many researchers have already utilized the PSO and ACO techniques efficiently for data mining to solve single- or multiple-objective problems.

3. **Management of Dynamical Data**
   The data encountered in network usage increases at every instance, and the real-time data information quickly changes with respect to the time. Handling the data from above cases using the analytical algorithms needs to be done swiftly. The dynamic data may be the result of non-stationary environments that dynamically changes in certain time. The swarm nintelligence was widely adopted by the big data concepts to address the issues that arise because of the wide range of uncertainties. Uncertain problems arise because of the following problems.

1. The data may be uncertain because of the presence of noise or the fitness function may not suit the conditions applied.

2. The parameters and design variables may change because of environmental parameters or there may be other the optimization solutions proposed for the data mining that should also keep kept in mind for future needs during designing.

3. The fitness function proposed may be or not be suitable for handling the data. All parameters must be considered clearly.

Applications:
Wireless sensor networks
There are many modern applications (temperature monitoring, habitat monitoring, etc.) for wireless sensor networks (WSN) in the monitoring and control area that are dependent on WSN. The WSN consists of hundreds of sensors distributed over an area, which sense the necessary information and communicate this information to the destination at regular intervals. The WSN jointly with the IoT is used for control applications. The combination of WSN and IoT can deal with many modern real-time problems. Massive data is generated from the large number of the nodes of the network. The important goal is to give the speediest possible solution toward beneficial information. Swarm intelligence is one of the smartest solutions for the effective handling of huge data and making it into useful information (Hadi, Lawey, El-Gorashi, & Elmirghani, 2018).

Intelligent transport systems
As the urban population is increasing every day in many cities, traffic is increasing everywhere, so the transportation system is facing many problems, such as weather conditions, an increasing number of vehicles, crashes, and traffic info variations in real time. Controlling traffic is a major issue because the amount of data generated is very high. Intelligent transport systems, with of help of swarm-intelligence-based big data analysis, can ensure intelligent transportation systems by controlling traffic jams, reducing the environmental pollution, etc. (Wang, Li, Zhou, Wang, & Nedjah, 2016).

9.6 Swarm Intelligence in Artificial Intelligence (AI)

AI deals with systems that work autonomously with their own thinking capacity. But many people who deal with AI are discussing its pros and cons, because AI may provide advantages at the present time in terms of technology, but they must also should deal with its disadvantages, such as AI's overtaking the decisions of human beings and operating technology autonomously in the future (Kalantari, Kamsin, Shamshirband, Gani, Rokny, & Chronopoulos, 2018).

9.6.1 Imitation of Birds, Fish, and Bees

Many organisms, such as fish, ants, and flocks of birds, in natural eco-systems perform their tasks by forming groups (swarms) and do all their work by the mutual interaction between swarm members. They strengthen their swarm intelligence by forming droves, schools, shoals, and colonies. Among many uncountable species, ecology shows us that social creatures, working as organized integrated unified systems, can do better than the huge majority of individual members when making decisions and resolving problems. Researchers and scientists define this as *"swarm intelligence,"* and it supports the old proverb *many minds are better than one* (Rosenberg, Pescetelli, & Willcox, 2017).

9.6.2 Human Intelligence vs. Swarm Intelligence

The human beings were not gifted with swarm intelligence by nature. Human beings are deficient in the subtle connections that other species were gifted with as they form a tight feedback-loops between members. Bees practice high-speed vibrations. The birds spot the motions propagating through the group. Fish sense tremors in the water around them. But with modern communication and networking technology, human beings are connected across the world. It is very important to rearrange this connection into real-time systems with closed-loop feedback among members (Rosenberg, 2016).

9.6.3 Swarm Intelligence and Artificial Intelligence

Human-based swarm technology is possible these days with modern communication technology. The combination of human-based swarm intelligence and artificial intelligence can emerge as a single *emergent intelligence.*

The present chapter deals with artificial swarm intelligence (ASI). ASI deals with the building the intelligent systems with humans in the loop. ASI deals the computational algorithms that are wise and creative, but with the intuition of the real people. There some are some real-time examples that deal with the ASI in daily activities. Enswarm, a startup company, uses ASI for decisions on employment, recruitment, and interview purposes. The startup Swarm Fund has modeled cryptocurrencies like Bitcoin using ASI technology, for fundraising (Illias & Chai, 2017).

The knowledge of human beings permits thinking only to a certain extent. Human intelligence does not show the ability for the formation of closed-loop swarms. Human-intelligence-based swarm intelligence is possible only with modern communication and networking technologies, and as these technologies are filling a void, this is possible only with efficient technology. Human intelligence combined with AI in combination with a vast communication and networking infrastructure embedded with new software allows the growing of an artificial swarm. This artificial swarm works efficiently with intelligent amplifications and can reach to good decisions similarly to the natural swarm. This is an exact example of hardware and software codesign creating efficient human–machine systems. The most important application using ASI is advanced medical diagnosis. This merges the knowledge of the many doctors and generates a most efficient solution for the for the emergent diagnosis that outdoes the experimental wisdom of the practitioner.

The advanced medical diagnosis performed by RAND Corporation and Humboldt-University of Berlin, a computational group of radiologists outperformed single practitioners in observing mammograms, decreasing false negatives and false positives. There was a similar study conducted by the Cleveland Clinic and John Carroll University in which a group of 12 radiologists analyzed skeletal abnormalities. The knowledge of the group of radiologists in the form of a computational collective has produced a significantly advanced rate of precise diagnosis compared to a single practitioner (Fortier, Sheppard, & Pillai, 2012).

9.7 Swarm Intelligence and the Internet of Things (IoT)

The most important functionality of the Internet is to attain information and distribute it among people. The IoT is more important for the relation between persons and things, machine to machine, and their connections to the network so as to recognize, manage, and control the system. IoT connects the various devices that belong to the physical world to an information-technology-based system. It utilizes the existing telecommunication network, Internet, and industry-oriented network, and adds some additional network capabilities to bring about the desired outputs. The IoT deals with sensing parameters, systematic transmission, and intelligent processing (Sangeetha, Bharathi, Ganesh, Radhakrishnan, 2018).

The performance of an IoT system depends on the embedded processing and its power consumption, interoperability, and security. The functioning and power efficiency are very important at the destination node level. The other requisites, such as security and interoperability, can have a direct influence on the end-user experience. The swarm technology ideologies are being accepted by IoT. The communication network is built for effective communication between many devices without losing the reliability of the data. In a communication network, there are many gateways. Each gateway is connected to many devices and can interface with many devices. All these devices are connected to achieve a great task collectively. It is the combined effort of all the nodes in the network that does every task (Ahmed & Rani, 2018).

Swarm intelligence is observed from nature and is based on the collective efforts of simple creatures like ants, bees, and termites. Massive heaps are built by termites. Bees build enormous stores effectively without loss. The modern applications inspired by the swarms of the bees building the hives can be used by the data networking technology. The knowledge obtained from thousands of very simple devices can be aggregated. The summation of the data can benefit specific pieces of the apparatus. The system can report problems in the network as the whole, and it can have future relevance to the Internet of Things. IoT is a network of simple devices and generates huge data; the collective effort of every individual in the network can address many challenges in terms of intelligence (Sood, Sandhu, Singla, & Chang, 2017). Figure 9.4. represents the architecture of the Internet of Things.

Any device in the network can interface to any device and support expansion because swarm intelligence was opted for in the data network-ing. No device in the network can be a private physical device. Instead, it can connect to a various number of dissimilar physical devices, such that each will contribute to its boundaries and add processing capabilities to the collective. Together, these individual devices can then be considered as a single entity in terms of architecture and function. This approach solves

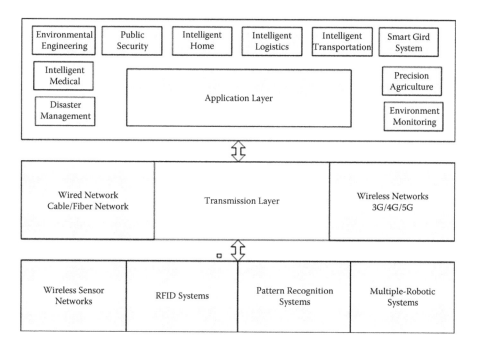

FIGURE 9.4
The architecture of the Internet of Things.

the scalability problem. Doing this results in an assessable reduction in the total cost of ownership. If you observe the system as a beekeeper, you can observe the following points belonging to IoT: devices capable of information gathering, data processing, inter-device information sharing (Sangeetha et al., 2018).

Learn from past mistakes; as exciting as IoT-device swarm intelligence seems, many technologists feel we have not learned from past mistakes when it comes to emergent technologies. There is a need to think about security issues that will occur after the technology becomes popular. Cloud computing and all the after-the-fact effort required to secure the technology is a case in point (Ahmed & Rani, 2018). Figure 9.5 represents the collective intelligence for the IoT using swarm intelligence.

There is a need to address the following (Ahmed & Rani, 2018; Sangeetha, 2018; Sood et al., 2017; Kim, Ramos, & Mohammed 2017):

- **Device authentication/identification**: Identifying and authenticating IoT devices comprising the swarm.
- **Confidentiality and integrity**: Protecting the data and allowing only authorized methods of manipulating the data.
- **Service availability**: Allowing only authorized personnel to have access, and keeping access from being maliciously denied.
- **There is a need for security at the end nodes**, where functioning and low power prerequisites make security difficult.
- **Increasing complication of the end node**: The more than 100 node networks like MANET and WBAN are now using wireless data communication, opening up new susceptibilities to data privacy.

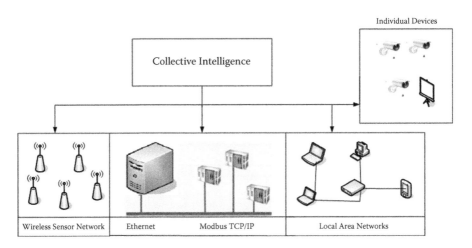

FIGURE 9.5
Swarm-based collective intelligence for IoT.

9.8 Conclusion

This chapter has discussed the role of swarm intelligence in various advanced engineering areas. Inherent properties of swarm intelligence as observed in nature include emergent behaviour, massive system scalability, intelligence from low complexity, autonomy, local interactions, and communication through the environment. These properties are very desirable for many types for many modern applications. Swarm-intelligence-based methods hold great ability for solving many problems. It is very clear that the social insects forming natural systems in the form of swarms can provide a high vision for designing algorithms and artificial problem solving problems. The swarm-based approaches are promising when a system is very complex, dynamic, and overloaded with information. When artificial-intelligence-based information is increasing and when complexity cannot be understood by human beings, swarm intelligence is found appealing by researchers. Swarm intelligence offers flexibility, decentralized control, self-organization, and robustness. Swarm intelligence has defined a new approach to intelligent systems with a focus on autonomy, emergence, centralization, and distributed functioning replacing control and preprogramming. The final result will be the accomplishing of very complex forms of social behaviour and self-actualization of a number of optimization and other tasks. The number of successful applications is exponentially growing in combinatorial optimization, communication networks, and robotics that employ swarm intelligence. The examples discussed in this chapter provide illustrations of these features. Swarm intelligence, however, is a new and exploratory field and much work remains to be done.

References

Abraham, A. & Ramos, V. (2003). Web usage mining using artificial ant colony clustering and genetic programming, *IEEE Congress on Evolutionary Computation (CEC2003), Australia*, IEEE Press, ISBN 0780378040, 1384–1391.

Ahmed, S. H. & Rani, S. (2018). A hybrid approach, smart street use case and future aspects for Internet of Things in smart cities. *Future Generation Computer Systems Part 3*, 79, 941–951. ISSN 0167-739X. https://doi.org/10.1016/j.future.2017.08.054.

Angiulli, F. (2018). Data mining: Outlier detection In *Reference Module in Life Sciences*. Elsevier, ISBN 9780128096338 Retrieved from https://doi.org/10.1016/B978-0-12-809633-8.20386-5. Amsterdam, Netherlands.

Bastos-Filho, C. J. A., Schuler, W. H., Oliveira, A. L. I., & Vitorino, L. N. (2008). Routing algorithm based on swarm intelligence and hopfield neural network applied to communication networks. *Electronics Letters*, 44(16), 995–997.

Bayındır, L. (2016). A review of swarm robotics tasks. *Neurocomputing*, 172, 292–321. ISSN 0925-2312. Retrieved from https://doi.org/10.1016/j.neucom.2015.05.116.

Bonidia, R. Parmezan, Brancher, J. Duilio and Busto, R. Marques. "Data Mining in Sports: A Systematic Review," in IEEE Latin America Transactions, vol. 16, no. 1, pp. 232-239, Jan. 2018. doi: 10.1109/TLA.2018.8291478

Borenstein, J., Everett, H. R., Feng, L., & Wehe, D. (1997). Mobile robot positioning: Sensors and techniques. *Journal of Intelligent and Robotic Systems, 14*(4), 231–249.

Çelik, F., Zengin, A., & Tuncel, S. (2010). A survey on swarm intelligence-based routing protocols in wireless sensor networks. *International Journal of the Physical Sciences, 5*(14), 2118–2126.

Cheng, Y., Chen, K., Sun, H., Zhang, Y., & Tao, F. (2017). Data and knowledge mining with big data towards smart production. *Journal of Industrial Information Integration*, ISSN 2452-414X. Retrieved from https://doi.org/10.1016/j.jii.2017.08.001

Cheng, S., Shi, Y., Qin, Q., & Bai, R. (2013). Swarm intelligence in big data analytics. In: H Yin, et al. (eds) *Intelligent Data Engineering and Automated Learning – IDEAL 2013, IDEAL 2013. Lecture notes in computer science,* Vol. 8206. Springer.

Cheng, S., Zhang, Q., & Qin, Q. (2016). Big data analytics with swarm intelligence. *Industrial Management & Data Systems,* 116(4), 646–666. Retrieved from https://doi.org/10.1108/IMDS-06-2015-0222.

Cortes, J., Martínez, S., Karatas, T., & Bullo, F. (2004). Coverage control for mobile sensing networks. *IEEE Transactions on Robotics and Automation,* 20(2), 243–255.

Deneubourg, J. L., Goss, S., Franks, N., Franks, A. S., Detrain, C., & Chretien, L. (1991). The dynamics of collective sorting: robot-like ants and ant-like robots. *Proceedings of the First International Conference on Simulation of Adaptive Behavior: From Animals to Animats,* Cambridge, MA: MIT Press, pp. 1, 356–365.

Dorigo, M., Bonabeau, E., & Theraulaz, G. (2000). Ant algorithms and stigmergy. *Future Generation Computing Systems,* 16(8), 851–871.

Grushin, A., & Reggia, J. A. (2006). Stigmergic self-assembly of prespecified artificial structures in a constrained and continuous environment. *Integrated Computer-Aided Engineering,* 13(4), 289–312.

Fayyad, U., Piatetsky-Shapiro, G., & Smyth, P. (1996). From Data Mining to Knowledge Discovery in Databases, *AI Magazine,* 17(3).

Fortier, N., Sheppard, J. W., & Pillai, K. G. (2012). DOSI: training artificial neural networks using overlapping swarm intelligence with local credit assignment. *The 6th International Conference on Soft Computing and Intelligent Systems, and The 13th International Symposium on Advanced Intelligence Systems, Kobe,* 1420–1425. doi: 10.1109/SCIS-ISIS.2012.6505078.

Gui, T., Ma, C., Wang, F., & Wilkins, D. E. (2016). "Survey on swarm intelligence-based routing protocols for wireless sensor networks: An extensive study," *2016 IEEE International Conference on Industrial Technology (ICIT), Taipei,* 1944–1949. DOI: 10.1109/ICIT.2016.7475064.

Hadi, M. S, Lawey, A. Q, El-Gorashi, T. E. H., & Elmirghani, J. M. H (2018). Big data analytics for wireless and wired network design: A survey. *Computer Networks, 132,* pp. 180–199. ISSN 1389-1286. Retrieved from https://doi.org/10.1016/j.comnet.2018.01.016.

Handl, J., Knowles, J., & Dorigo, M (2006). Ant-based clustering and topographic mapping. *Artificial Life,* 12(1), pp. 35-61.

Hawick, K. A., James, H. A., Story, J. E., & Shepherd, R. G. (2002). *An Architecture for Swarm Robots* (DHCP-121). Wales: University of Wales.

Illias, H. A., & Chai, X. R. (2017). Transformer insulation diagnosis using hybrid artificial intelligence-modified particle swarm optimization. *2017 IEEE*

Conference on Energy Conversion (CENCON), Kuala Lumpur, 255–258. doi: 10.1109/CENCON.2017.8262494.

Juan, L., Chen, S., Chao, Z. (2007). Ant system based anycast routing in wireless sensor networks. *2007 International Conference on Wireless Communications, Networking and Mobile Computing, Shanghai,* 2420–2423.

Javan, M. R., Mokari, N., Alavi, F., & Rahmati, A. (2017, January). Resource allocation in decode-and-forward cooperative communication networks with limited rate feedback channel. *IEEE Transactions on Vehicular Technology, 66*(1), 256–267. 10.1109/TVT.2016.2550103.

Kalantari, A., Kamsin, A., Shamshirband, S., Gani, A., Hamid Alinejad-Rokny, A. T., & Chronopoulos, A. T. (2018). Computational intelligence approaches for classification of medical data: State-of-the-art, future challenges and research directions. *Neurocomputing, 276,* 2–22. ISSN 0925-2312. Retrieved from https://doi.org/ 10.1016/j.neucom.2017.01.126.

Karia, D. C., & Godbole, V. V. (2013). New approach for routing in mobile ad-hoc networks based on ant colony optimization with global positioning system. *IET Networks, 2*(3), 171–180. 10.1049/iet-net.2012.0087.

Kassabalidis, M. A., El-Sharkawi, R. J., Arabshahi, M.P., & Gray, A. A. (2001). Swarm intelligence for routing in communication networks *Global Telecommunications Conference (GLOBECOM '01) 2001.* IEEE, San Antonio, TX (Vol. 6, pp. 3613–3617). doi: 10.1109/GLOCOM.2001.966355.

Kaur, S., & Sharma, S. (2017). Performance evaluation of artificial bee colony and compressive sensing based energy efficient protocol for WSNs. *13th International Conference on Signal-Image Technology & Internet-Based Systems (SITIS), Jaipur, India,* 91–96.

Kim, T.-H., Ramos, C., & Mohammed, S. (2017). Smart City and IoT. *Future Generation Computer Systems, 76,* 159–162. ISSN 0167-739X. Retrieved from https://doi. org/10.1016/j.future.2017.03.034.

Sangeetha, Lakshmi, A., Bharathi, N., Ganesh, Balaji, & Radhakrishnan, T. K. (2018). Particle swarm optimization tuned cascade control system in an Internet of Things (IoT) environment. *Measurement, 117,* 80–89. ISSN 0263-2241. Retrieved from https://doi.org/10.1016/j.measurement.2017.12.014

Ling, L. & Feng, Z (2013). Partner selection of agile supply chain based on MAX-MIN ant system. *6th International Conference on Information Management, Innovation Management and Industrial Engineering, Xi'an,* 531–534.

Loganathan, J., Logitha, X. X., & Veronica, A. (2015). Distributed resource management scheme using enhanced artificial bee-colony in P2P. *2nd International Conference on Electronics and Communication Systems (ICECS), Coimbatore,* 1035–1039.

Ma, M., Sun, C., & Chen, X. (2017). Discriminative deep belief networks with ant colony optimization for health status assessment of machine. *IEEE Transactions on Instrumentation and Measurement, 66*(12), 3115–3125. 10.1109/ TIM.2017.2735661.

Paone, M., Paladina, L., Bruneo, D., & Puliafito, A. (2007). A swarm-based routing protocol for wireless sensor networks *Sixth IEEE International Symposium on Network Computing and Applications (NCA 2007), Cambridge, MA,* 265–268.

Parmezan Bonidia, R., Duilio Brancher, J., & Marques Busto, R. (2018). Data mining in sports: A systematic review. In *IEEE Latin America Transactions* (Vol. 16, No. 1, pp. 232–239). doi: 10.1109/TLA.2018.8291478.

Payton, D., Estkowski, R., & Howard, M. (2005). Pheromone robotics and the logic of virtual pheromones. In *Swarm robotics Lecture notes in computer science* (Vol. 3342, pp. 45–57, Springer. Berlin, Germany.

Ramírez-Gallego, S., Fernández, A., García, S., Chen, M., & Herrera, F. (2018). Big data: Tutorial and guidelines on information and process fusion for analytics algorithms with MapReduce. *Information Fusion, 42*, 51–61. ISSN 1566-2535. Retrieved from https://doi.org/10.1016/j.inffus.2017.10.001.

Ramesh, S., Praveen, R., Indira, R., & Kumar, P. G. (2013). Bee routing protocol for intermittently connected mobile networks. *International Conference on Recent Trends in Information Technology (ICRTIT), Chennai*, 175–181.

Ranjbar-Sahraei, B., Weiss, G., & Nakisaee, A. (2012). A multi-robot coverage approach based on stigmergic communication. In *Multiagent System Technologies Lecture Notes in Computer Science* (Vol. 7598, pp. 126e38), Springer.

Rosenberg, L. (2016). Artificial swarm intelligence vs human experts *2016 International Joint Conference on Neural Networks (IJCNN), Vancouver, BC*, 2547–2551. doi: 10.1109/IJCNN.2016.7727517.

Rosenberg, L., Pescetelli, N., & Willcox, G. (2017) Artificial swarm intelligence amplifies accuracy when predicting financial markets *2017 IEEE 8th Annual Ubiquitous Computing, Electronics and Mobile Communication Conference (UEMCON), New York City, NY*, pp. 58–62.doi: 10.1109/UEMCON.2017.8248984.

Senanayake, M., Senthooran, I., Barca, J. C., Chung, H, Kamruzzaman, J., & Murshed, M. (2016). Search and tracking algorithms for swarms of robots: A survey. *Robotics and Autonomous Systems Part B, 75*, 422–434. ISSN 0921-8890. Retrieved from https://doi.org/10.1016/j.robot.2015.08.010

Shenoy, M. V., & Anupama, K. R. (2018). Swarm-Sync: A distributed global time synchronization framework for swarm robotic systems. *Pervasive and Mobile Computing, 44*, 1–30, ISSN 1574-1192. Retrieved from https://doi.org/10.1016/j.pmcj.2018.01.002

Singh, B. & Aggarwal, P. (2017). Using hybrid ACO/PSO enhance NN RZ leach for mobile sink in WSN's *8th IEEE Annual Information Technology, Electronics and Mobile Communication Conference (IEMCON), Vancouver, BC*, 546–552.

Sood, S. K., Sandhu, R., Singla, K., & Chang, V. (2017). IoT, big data and HPC based smart flood management framework, *Sustainable Computing: Informatics and Systems*, ISSN 2210-5379, https://doi.org/10.1016/j.suscom.2017.12.001.

Suárez, P., Iglesias, A., & Gálvez, A. (2018, February). Make robots be bats: Specializing robotic swarms to the bat algorithm, swarm, and evolutionary computation, ISSN 2210-6502 Retrieved from https://doi.org/10.1016/j.swevo.2018.01.005.

Tatomir, B., Rothkrantz, L. J. M., & Suson, A. C. (2009). Travel time prediction for dynamic routing using ant based control. *Proceedings of the 2009 Winter Simulation Conference (WSC), Austin, TX*, 1069–1078. DOI: 10.1109/WSC.2009.5429648.

Urso, A., Fiannaca, A., La Rosa, M., Ravì, V., & Rizzo, R. (2018). Data mining: Classification and prediction. In *Reference Module in Life Sciences*, Elsevier, ISBN 9780128096338. Retrieved from https://doi.org/10.1016/B978-0-12-809633-8.20461-5.

Vadim, K. (2018). Overview of different approaches to solving problems of data mining. *Procedia Computer Science*, Vol. 123, 234–239. ISSN 1877-0509. Retrieved from https://doi.org/10.1016/j.procs.2018.01.036.

Wang, C., Li, X., Zhou, X., Wang, A., & Nedjah, N. (2016). Soft computing in big data intelligent transportation systems. *Applied Soft Computing, 38* 1099–1108. ISSN 1568-4946. Retrieved from https://doi.org/10.1016/j.asoc.2015.06.006.

Wei, J., & Wang, L. (2017). Big data analytics based optimization for enriched process planning: A methodology. *Procedia CIRP, 63,* 161–166. ISSN 2212-8271. Retrieved from https://doi.org/10.1016/j.procir.2017.03.090.

Zhang, Y., Kuhn, L. D., & Fromherz, M. P. (2004, September). Improvements on ant routing for sensor networks. In *International Workshop on Ant Colony Optimization and Swarm Intelligence* (pp. 154-165). Springer, Berlin, Heidelberg.

Zheng, F., Zecchin, A. C., Newman, J. P., Maier, H. R., & Dandy, G. C. (2017). An adaptive convergence-trajectory controlled ant colony optimization algorithm with application to water distribution system design problems. *IEEE Transactions on Evolutionary Computation, 21*(5), 773–791. 10.1109/TEVC.2017.2682899.

Index